SolidWorks for the Sheet Metal Guy
Course 1: Part Creation

Joe Bucalo
Neil Bucalo

www.SheetMetalGuy.com

SolidWorks for the Sheet Metal Guy
Course 1: Part Creation
Published by
Sheet Metal Guy, LLC
P O Box 498283
Cincinnati, OH 45249
www.SheetMetalGuy.com

Every effort has been made to ensure that all information contained within this book is complete and accurate. However, Sheet Metal Guy assumes no responsibility for the use of said information, nor any infringement of the intellectual property rights of third parties which would result from such use.

Please visit our website at www.SheetMetalGuy.com.

Manufactured in the United States of America.

All technical illustrations and CAD models in this book were produced using SolidWorks 2006. The commands throughout this book have been tested for compatibility with SolidWorks 2007.

ISBN 978-0-9795666-1-5

About the Authors

Joe Bucalo is the founder and President of Applied Production, Inc. He has over 30 years experience in the sheet metal industry, working with and developing software products to change the manufacturing world. When CAD was just beginning on the PC, Joe played a major role in the development of ProFold, the first truly automatic 3D sheet metal unfolding program. When most people were using a wall chart of bend deductions, Joe was promoting the use of the K-factor for more accurate flat patterns.

Joe later paved the way for graphics based sheet metal CAM when he introduced ProFab, which allows the direct transfer of geometry from CAD to CAM. ProFab was the first CAM program in the sheet metal industry to include an automatic tool selection routine.

Joe continues to work with clients to solve their design and manufacturing problems. He has a thorough knowledge of the most popular CAD programs and understands the issues faced by sheet metal manufacturers.

Neil Bucalo has a diverse background, including mechanical design engineering, CAD/CAM support and training, engineering consulting, web development, and technical writing. Neil started his career in support and training of the CADKEY software at Computer Aided Technology, Inc. He then moved forward as a Certified SolidWorks Support Technician.

Upon moving to the Cincinnati area, Neil joined Applied Production, Inc., a SolidWorks Solution Partner, where he has provided customer support and written several user training documents. He also created and served as Editor of the CKD Magazine, dedicated to users of the CADKEY software.

Neil is a CAD expert, having many years of experience using numerous CAD systems, including AutoCAD, CADKEY/KeyCreator, Solid Edge, and of course SolidWorks.

For the record, Neil is Joe's nephew.

Tell Us What You Think!

As the reader of this book, you are our most important critic. We value your opinion and want to know how we are doing, good or bad. If you feel we missed something or could have done a better job, let us know. Also, if there are other areas of SolidWorks you feel need more explanation, tell us. We may be able to help.

You can email us at **books@SheetMetalGuy.com** to let us know what you did or didn't like about this book – as well as what we can do to make our books better.

When you write, please be sure to include the book's title as well as your name and contact information. We will carefully review your comments and share them with those whom helped make this book possible.

Table of Contents

Introduction

Since the day Applied Production, Inc. first released ProFold to the sheet metal world, phone calls have been received from customers struggling to get the CAD geometry right and to calculate the correct flat pattern. In the beginning, the problem was working in 3D (not front, top, and side views, but real 3D modeling with wireframe). Today, it's solid modeling.

The CAD world offers a wide array of training materials and tutorials, but all are geared to machining and mold making. While the CAD software includes a sheet metal toolkit and unfolding capabilities, information about the use of these features is scarce. And worse yet, the typical CAD reseller lacks real sheet metal knowledge. They use demonstrations and modeling techniques which create a machined part, and then they want to "thin wall shell" it, as they say, followed by "ripping the corners". And poof, it's a sheet metal part, or so they think. Common phrases include words like surface and extrude. These are not the words of a sheet metal person.

The purpose of this series of books is to teach you how the sheet metal tools work, demonstrating how to apply parametrics to the part so that changes in the material thickness do not change the finished part. The most important thing is to talk in terms that sheet metal people know and understand.

Conventions Used in this Book

It is assumed that you have a working knowledge of SolidWorks and the menu structure. You may want to open SolidWorks and in the "Help" menu, go through the **Online Tutorial**. In the first few chapters, we show the full CommandManager to help you learn what icons to select. Later chapters show only the icon to be selected. Dialog boxes, toolbars, and icons are shown in the book. When several icons appear in a dialog box, the one which you should select is circled in the picture in the book. A circle will not appear on your SolidWorks screen.

Setting the Toolbars to Match the Book

The CommandManager is a context-sensitive toolbar that dynamically updates based on the toolbar you want to access. By default, it has toolbars embedded in it based on the document type.

When you click a button in the control area, the CommandManager updates to show that toolbar. For example, if you click **Sketch** in the control area, the "Sketch" toolbar appears in the CommandManager.

Use the Command Manager to access toolbar buttons in a central location and to save space for the graphics area.

To access the CommandManager, first open a new document. To do this, select the **New** icon in the "Standard" toolbar, or pull down the "File" menu and select **New**.

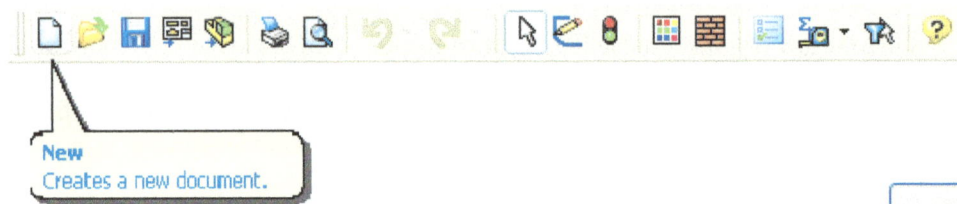

The **New SolidWorks Document** dialog box appears.

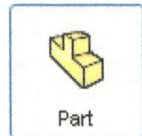

Select **Part** and then select **OK**. A new part window appears.

Pull down the "Tools" menu and select **Customize**.

In the **Customize** dialog box on the **Toolbars** tab, make sure the **Enable CommandManager** check box is checked. If it is not, select the check box to check it. Then, select the **OK** button.

To make it easier for you to follow along and find the commands described in this book, you will want to make certain that the CommandManager is the same as displayed in the book.

To set up the CommandManager, move the cursor over the CommandManager and click the right mouse button. In the menu, make sure that **Show Description** is checked. Then, select **Customize Command Manager** as shown.

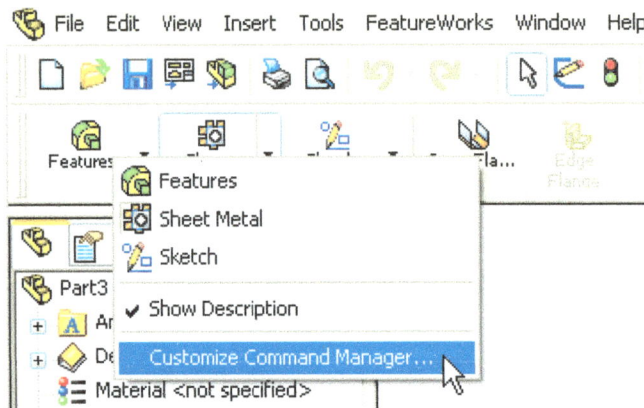

A long menu will appear, as shown to the right. Check **Features**, **Sheet Metal**, and **Sketch**. Make sure that all the others are unchecked. To accept your selections, simply click the left mouse button anywhere in the graphics area.

Chapter 1

Top Hat

For the purpose of this book, sheet metal parts will be created in SolidWorks beginning with a Base Flange. This may be drawn as the outline of the flange shape or the cross section of the part. Here you will use the cross section method to create a Top Hat. The two outside flanges will be equal so that changing one will update the other.

While it is easier to create the cross section as the starting point for many parts, modifying the flanges can be more difficult with this method. Therefore most of your parts will be created from a rectangular base flange.

This first part also features a rectangular cutout and four mounting holes just to get you acquainted with these commands. More detail about these commands is included in Course 2.

By properly placing the key dimensions and using an equation to account for the material thickness, you can change the height, width, and length of the part while maintaining a constant shape. The material thickness may also be changed without affecting the part shape or size.

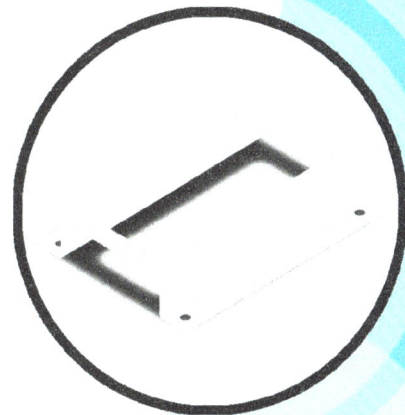

Creating a New Part Document

To begin this chapter, create a new part document in SolidWorks.

Select the **New** icon in the "Standard" toolbar, or pull down the "File" menu and select **New**.

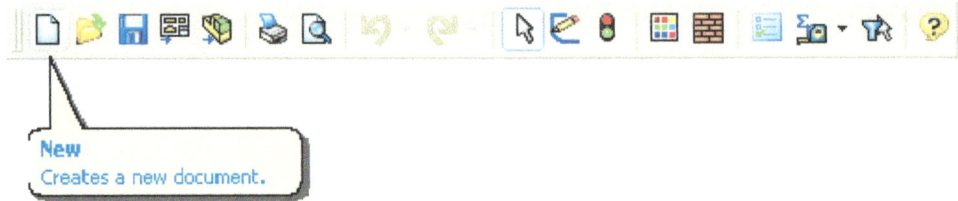

New
Creates a new document.

The **New SolidWorks Document** dialog box appears.

Select **Part** and then select **OK**. A new part window appears.

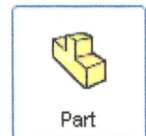

Part

The Base Flange

The first feature that you will create in this part is a base flange.

To do this, select the **Sheet Metal** icon in the control area of the CommandManager. Then, select the **Base-Flange/Tab** icon from the toolbar, or pull down the "Insert" menu and select **Sheet Metal – Base Flange**.

Features   Sheet Metal   Sketch   Base-Fla...   Edge Flange   Miter Flange

Base-Flange/Tab
Creates a sheet metal part or adds
material to an existing sheet metal part.

Front Plane

Top Plane

Front Plane

Right Plane

The **Front**, **Top**, and **Right** planes appear, and the cursor changes to

You are prompted in the PropertyManager, on the left hand side of the screen, to select a plane on which to sketch the part cross-section. As you move the cursor over a plane, the border of the plane is highlighted.

Move the cursor so that the **Front Plane** is highlighted. Click the left mouse button to select the **Front Plane**.

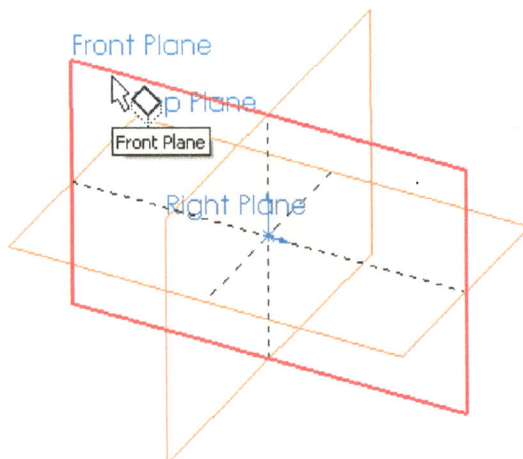

The display changes so that the **Front Plane** is facing you and SolidWorks opens a new sketch on this plane. The CommandManager dynamically updates to show the "Sketch" toolbar.

Select the **Line** icon in the CommandManager, or pull down the "Tools" menu and select **Sketch Entities – Line**.

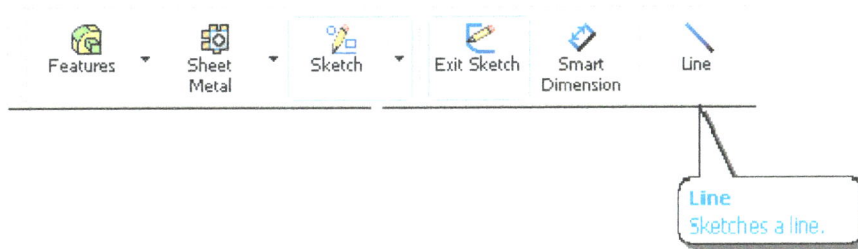

Using the **Line** tool, you are going to create the end profile of the Top Hat part. The sketch will then be given a thickness and a depth. The final sketch is shown below.

Move the cursor to the origin marker.

A large red dot appears when you are there.

The cursor is on the origin when the cursor changes from   to

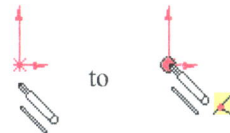

Pick the origin by clicking the left mouse button.

Move the cursor to the right to create a horizontal line. As you move the cursor, it displays a preview of the line. Do not concern yourself with the length as you will use the **Smart Dimension** tool to dimension the sketch to the desired size. But make certain that you see the **Horizontal** relation symbol next to the cursor indicating that the line is horizontal.

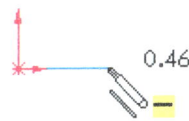

0.46 To complete the line, click the left mouse button.   0.81

Now, move the cursor up from its current location to create a vertical line. A preview line is still attached to the cursor from the end point of the previous line allowing you to continue to create a string of lines. This time, make certain you see the **Vertical** relation symbol next to the cursor indicating that the line is vertical.

Create five lines as shown below to sketch out the profile of the Top Hat. Take note of the dashed lines which appear on the screen when the cursor is in line with another line or end point. Once the fifth line is created, simply press the **Escape** key on the keyboard to exit out of the **Line** command.

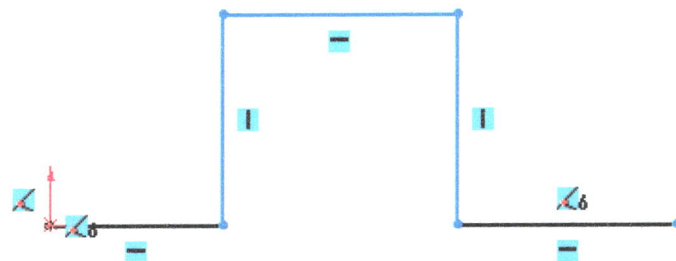

Adding Dimensions

In this section, the size of the part will be specified by adding dimensions. The SolidWorks software does not require that sketches are dimensioned before they are used to create features. However, for this chapter, dimensions will be added to fully define the sketch.

Select the **Smart Dimension** icon in the CommandManager, or pull down the "Tools" menu and select **Dimensions – Smart**.

Smart Dimension
Creates a dimension for one or more selected entities.

The cursor changes from to .

Select the top horizontal line of the part, and then move the cursor up to where you want to place the dimension and click the left mouse button.

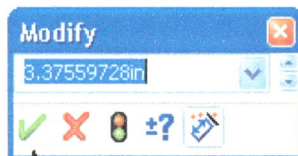

Save the current value and exit the dialog.

The **Modify** dialog box appears and the current dimension value is highlighted.

Type the number '**6**' and then select or press the **Enter** key on the keyboard. The sketch changes size to reflect the new dimension. The dimension value is now **6** inches.

Select the **Zoom to Fit** icon in the "View" toolbar, or pull down the "View" menu and select **Modify – Zoom to Fit**, so you can see the entire sketch and to center it in the graphics area.

Zoom to Fit
Zooms the model to fit the window.

Dimension the sketch by selecting the lines shown and modifying the dimension value.

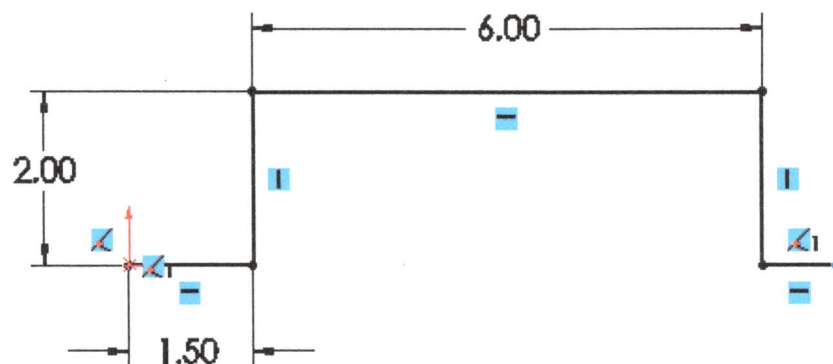

SolidWorks for the Sheet Metal Guy

If one of the legs flipped inside, don't get upset. It is really very easy to correct this. Start by pressing the **Escape** key on the keyboard. Then, move the cursor onto the end of the line you want to move. A red dot will appear on the endpoint. Press and hold the left mouse button, then drag the endpoint to the outside of the part. Again, don't worry about the length. You will fix that next.

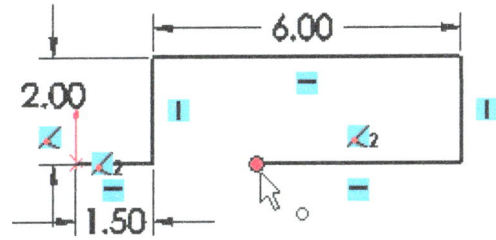

Add Relations

To make the two outside lines the same length, or equal, press the **Escape** key to clear any active functions or selected entities. Select the **Add Relation** icon in the CommandManager, or pull down the "Tools" menu and select **Relations – Add**.

Select the two small horizontal lines on the bottom of the profile.

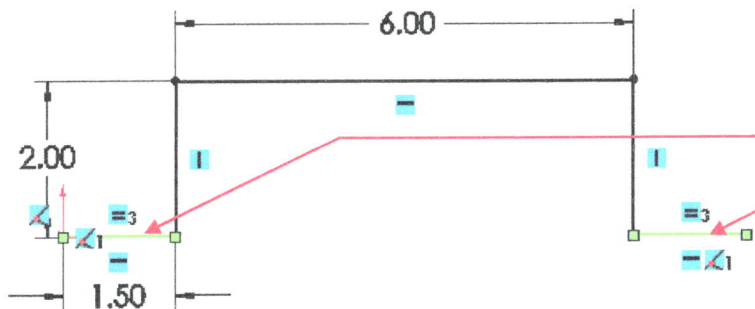

Next, select the **Equal** button in the **Add Relations** window of the **Properties** PropertyManager to make the length of the lines equal.

Select the green check mark button at the top of the **Properties** PropertyManager to accept the settings and update the geometry.

Select the **Exit Sketch** icon in the CommandManager or in the upper right corner of the graphics area.

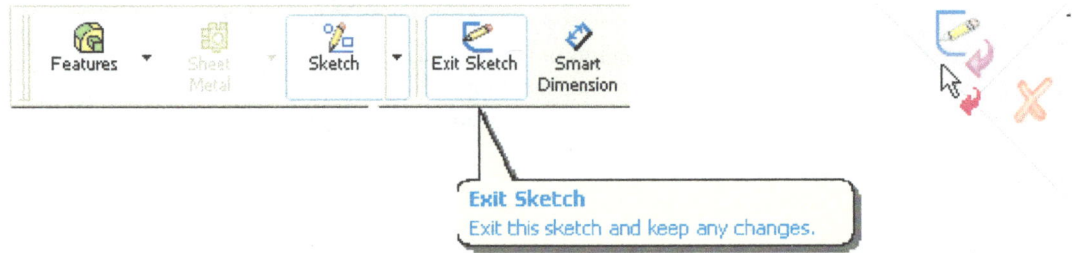

Completing the Base Flange

The **Base Flange** PropertyManager appears on the left side of the SolidWorks window, the graphics area changes to the standard **Trimetric** view, and a preview of the base flange appears. Note the material thickness appears below the sketch of the Top Hat.

In the **Base Flange** PropertyManager under **Direction 1**, set the **End Condition** to **Blind** and select the **Reverse Direction** button so that the preview of the part is behind the sketch. If you can't see these properties, click on the little triangle to the right of **Direction 1** in the **Base Flange** PropertyManager.

Set the **Depth** to '**12.00in**'.

Under **Sheet Metal Parameters**, set the **Thickness** to '**0.059in**'.

Set the **Bend Radius** to '**0.075in**'.

Select the green check mark button at the top of the **Base Flange** PropertyManager to accept the settings and create the part.

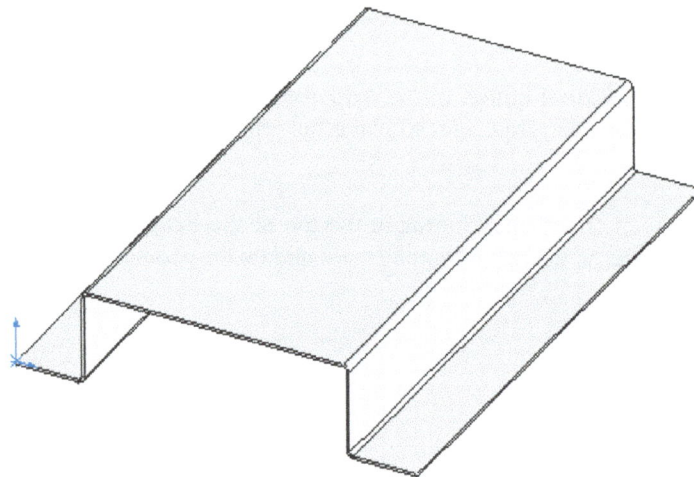

If the part is off the screen, select the **Zoom to Fit** icon in the "View" toolbar, or pull down the "View" menu and select **Modify – Zoom to Fit** so you can see the entire part and center it in the graphics area.

To see what you have created so far, in the FeatureManager design tree click the left mouse button on the plus sign to the left of **Base-Flange1**. Listed under **Base-Flange1** is **Sketch1**, which you used to create this feature. Immediately below **Sketch1** are the four bend areas of the part.

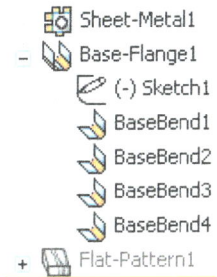

Creating a base flange defines the following features in the FeatureManager design tree: **Sheet-Metal1**, **Base-Flange1**, **BaseBend1**, **BaseBend2**, **BaseBend3**, **BaseBend4**, **Flat-Pattern1**.

- Sheet-Metal1
- Base-Flange1
  - (-) Sketch1
  - BaseBend1
  - BaseBend2
  - BaseBend3
  - BaseBend4
- Flat-Pattern1

Sheet-Metal1

Contains the default bend parameters including the material thickness, bend radius, bend allowance, and relief type. To edit any of these, right click on **Sheet-Metal1** and select **Edit Feature** from the menu.

Base-Flange1

Designates the first solid feature of the sheet metal part. To edit the Base-Flange parameters, right click on **Base-Flange1** and select **Edit Feature** from the menu.

BaseBend

Each bend area is represented by one of these features. Clicking on one of the **BaseBend** features, highlights the bend area of the part in the graphics area. To edit the **BaseBend** parameters, right click on the BaseBend you wish to edit and select **Edit Feature** from the menu. Here you can specify a bend radius or bend allowance different than the default.

Flat-Pattern1

This represents the unfolded part. The flat pattern is suppressed by default because the part is being created in its formed state. To unfold the part, right click **Flat-Pattern1** and select **Unsuppress**. When the Flat-Pattern feature is suppressed (your view is of the folded part), new features are automatically inserted above the Flat-Pattern feature in the FeatureManager design tree. When the Flat-Pattern feature is unsuppressed (your view is of the unfolded part), new features are added below it in the FeatureManager design tree and are not shown in the folded part. It is best to add features to the folded 3D part because of this. If you need to add tooling holes or embosses to your part for manufacturing purposes, you may want to add them to the unfolded part (below the Flat-Pattern feature). This makes them available when the part is flat and sent to your CNC programmer, yet they are not visible in the folded 3D model.

Show the Dimensions

Now that you have created the sheet metal part, you may want to display the dimensions.

In the FeatureManager design tree, right click on **Annotations** and select **Show Feature Dimensions**. Make sure that **Display Annotations** is also checked. You may drag and drop dimensions with the mouse to reposition them and make them easier to see.

If the part or the dimensions are off the screen, select the **Zoom In/Out** icon in the "View" toolbar, or pull down the "View" menu and select **Modify – Zoom In/Out**. Drag the cursor up (toward the top of the screen) to zoom in, or drag down to zoom out.

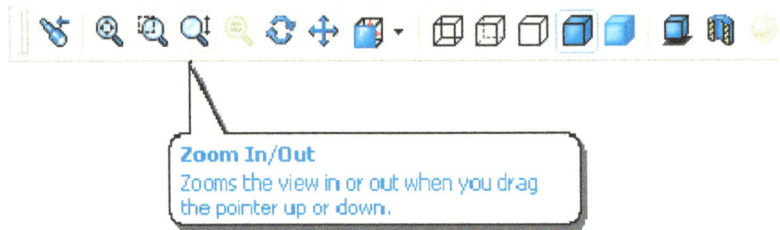

Zoom In/Out
Zooms the view in or out when you drag the pointer up or down.

If you use a three-button mouse, you can zoom by holding down **Shift** and dragging with the middle mouse button.

Using an Equation

In the bottom left corner of your graphics area, change the View orientation by selecting the pull down arrow and selecting **Front**.

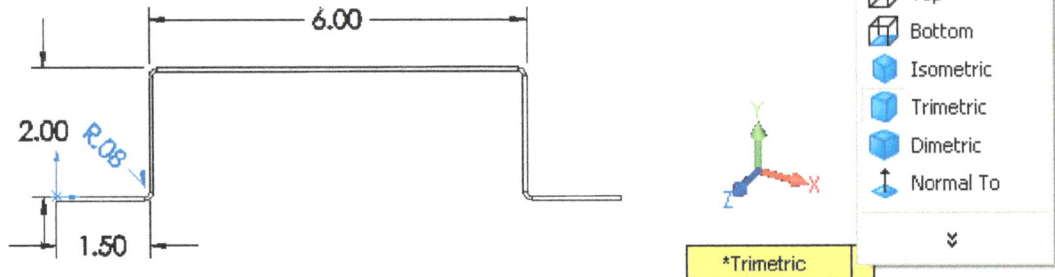

The overall height (outside to outside) of the part should be 2.00. Although the dimension reads 2.00, the dimension is measuring from the origin, not the bottom of the material thickness, as it should. To correct this dimension problem, an equation will be used. An equation is a mathematical relation between model dimensions.

Pull down the "Tools" menu and select **Equations**. When a sheet metal part is created, the variable "Thickness" is automatically created and linked to the thickness of the part.

In the **Equations** dialog box, select the **Add** button.

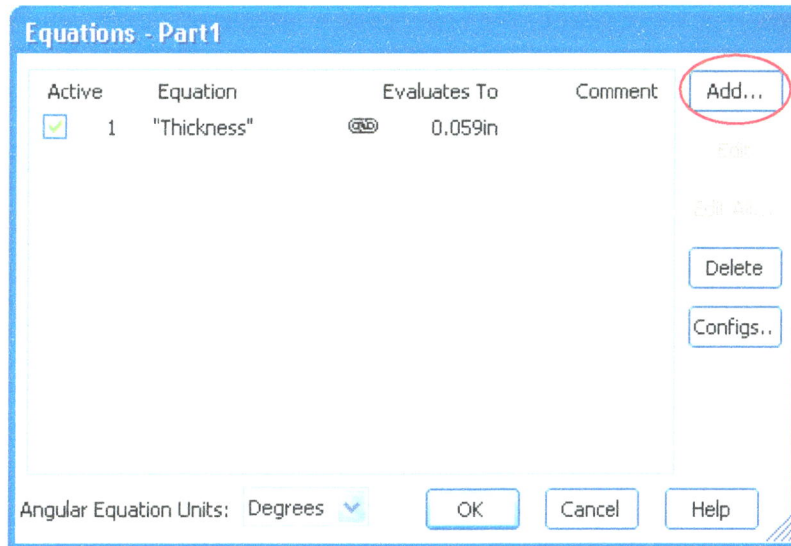

Select the **2.00** dimension in the graphics area. If the dialog boxes are covering the dimension, place the cursor on the blue border at the top of the dialog box, press and hold the left mouse button, and drag the dialog box out of the way. The variable name for this dimension will appear in the **Add Equation** dialog box.

Next, select the **Equals** button or press the '=' key on the keyboard.

Then, select the **2** button or press the '2' key on the keyboard.

Select the **Minus** button or press the '-' key on the keyboard.

Type the variable name '**Thickness**' to finish the equation.

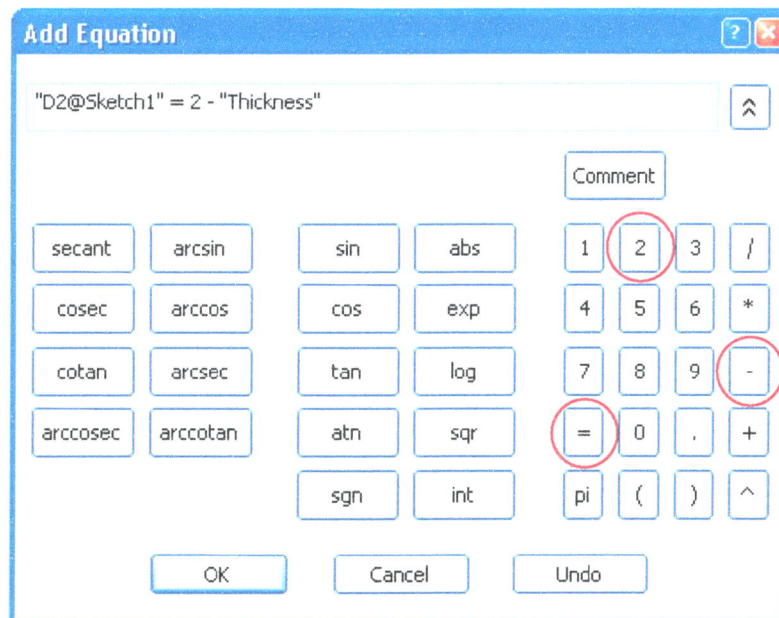

Finally, select the **OK** button to create the equation.

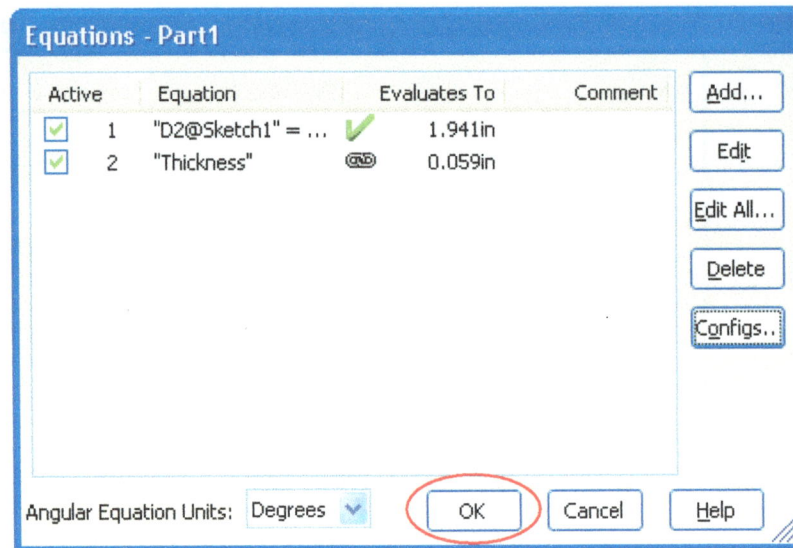

Select the **OK** button to exit the **Equations** dialog box.

The **2.00** dimension now reads Σ **1.94**.

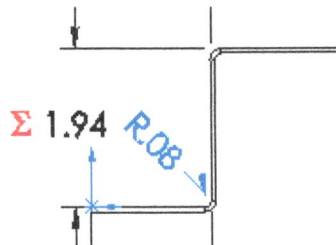

Select the **Rebuild** icon in the "Standard" toolbar, or press the keyboard shortcut **Ctrl+B**.

Rebuild
Rebuilds the part, assembly, or drawing.

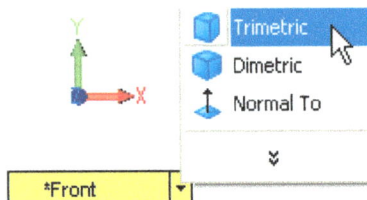

In the bottom left corner of your graphics area, change the View orientation by selecting the pull down arrow and selecting **Trimetric**. You can also display this list of views by pressing the spacebar on the keyboard.

Add a Cutout

Cut-Extrude is a method of putting a hole or cutout in the part. You are now going to create a large rectangular cutout on the top surface of the part. To do this, select the **Extruded Cut** icon from the "Sheet Metal" toolbar or pull down the "Insert" menu and select **Cut – Extrude**.

Extruded Cut
Cuts a solid model by extruding a sketched profile in one or two directions.

When prompted to select a planar face on which to sketch the feature cross-section, select the top face of the Top Hat as shown.

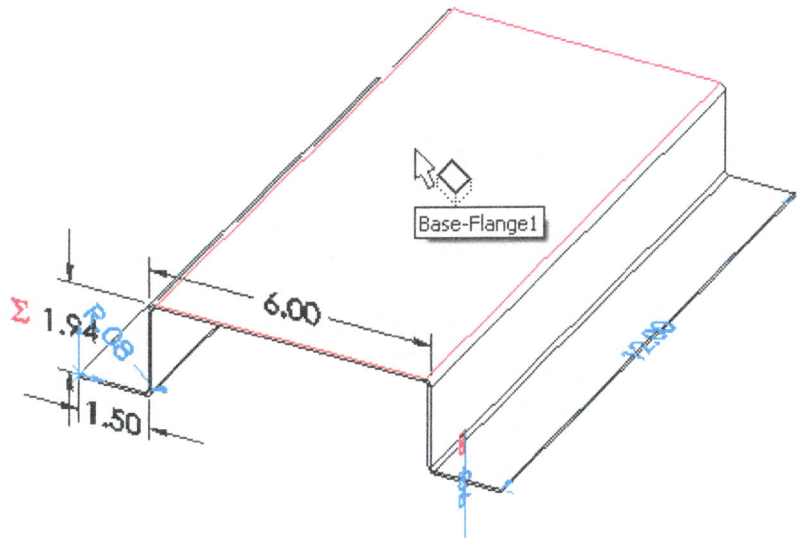

The CommandManager changes to the "Sketch" toolbar. Select the **Rectangle** icon in the CommandManager, or pull down the "Tools" menu and select **Sketch Entities – Rectangle**.

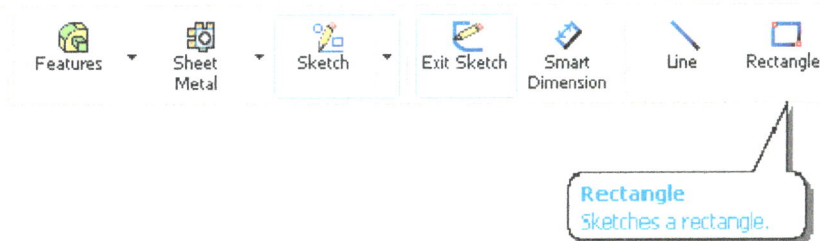

Rectangle
Sketches a rectangle.

Pick a starting location near the lower left hand corner of the top of the part. Move the cursor up and to the right to create a rectangle. As you move the cursor, an outline of the rectangle is displayed.

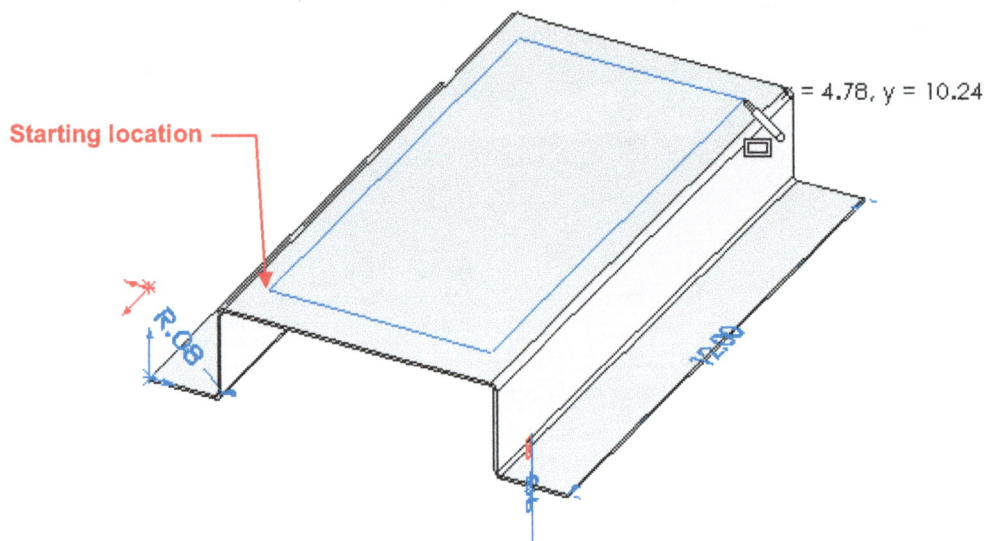

Starting location

= 4.78, y = 10.24

To complete the rectangle, click the left mouse button. The size of the rectangle does not matter. You will use the **Smart Dimension** tool to dimension the sketch so that each line is 1 inch from the edge of the part.

To do this, select the **Smart Dimension** icon in the CommandManager, or pull down the "Tools" menu and select **Dimensions – Smart**.

Features   Sheet Metal   Sketch   Exit Sketch   Smart Dimension

Smart Dimension
Creates a dimension for one or more selected entities.

The cursor changes from   to   .

Select the top sketch line and then select the parallel edge closest to the sketch line. Even though a dimension appears after you select the first line, go ahead and pick the second line. SolidWorks will magically change to dimension the distance between these two lines. Now move the cursor up and click where you want to place the dimension.

Select these two lines

In the **Modify** dialog box, enter the number '**1**' and then select ✔ or press the **Enter** key on the keyboard.

Do the same for the bottom sketch line, dimensioning it to be 1 inch from the edge of the part.

The left and right sketch lines are a little trickier. The dimension that is needed is from the outside of the material to the sketch line.

Select the outside bend line and the sketched line. After placing the dimension, type the number '**1.00**' in the **Modify** dialog box, and then press the **Enter** key on the keyboard.

Select these two lines

Do the same for the left sketch line, dimensioning it to be 1 inch from the edge line. You may need to rotate the part to make sure that you are selecting the correct line. Select the **Rotate View** icon in the "View" toolbar, or pull down the "View" menu and select **Modify – Rotate**.

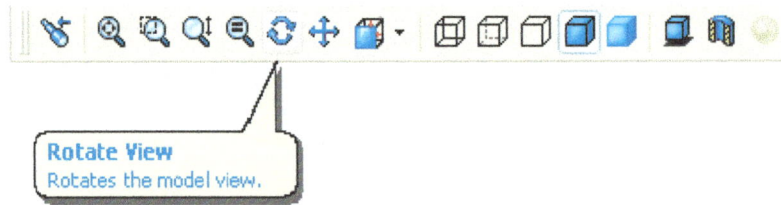

Rotate View
Rotates the model view.

The cursor will change to ⟳ . Drag the cursor to rotate the view.

You may also rotate the model view by holding down the middle mouse button and dragging the cursor or by using the arrow keys to rotate the view vertically or horizontally. (Hold down **Shift** to rotate in 90-degree increments. Hold down **Alt** and use the left-right arrow keys to rotate clockwise or counterclockwise.) If you use a three-button mouse, you can rotate the part by dragging with the middle mouse button. If you use a wheel-type mouse, you can also zoom in and out by spinning the wheel. While spinning the mouse wheel, you must keep the pointer on the area where you want to zoom

To return back to the active command you were using, simply select the **Rotate View** icon again.

Select the two lines as indicated and create the dimension.

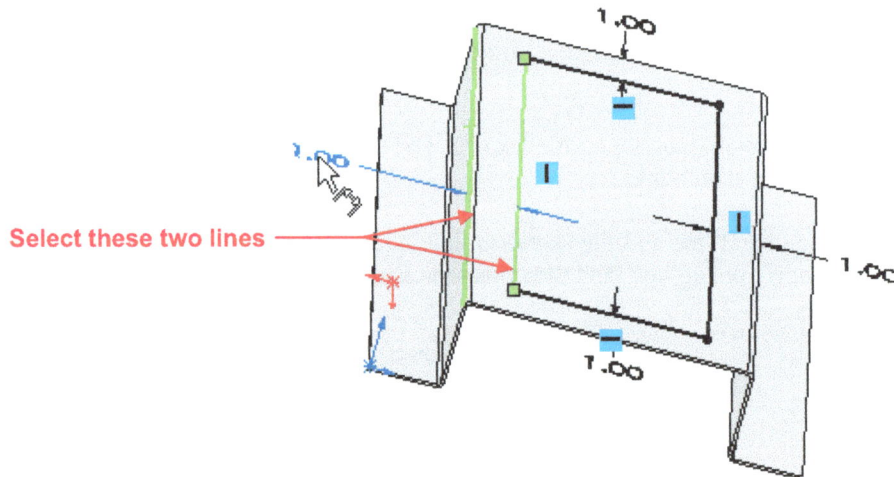

Select these two lines

In the bottom left corner of your graphics area, change the View orientation by selecting the pull down arrow and selecting **Trimetric**. You can also display this list of views by pressing the spacebar on the keyboard.

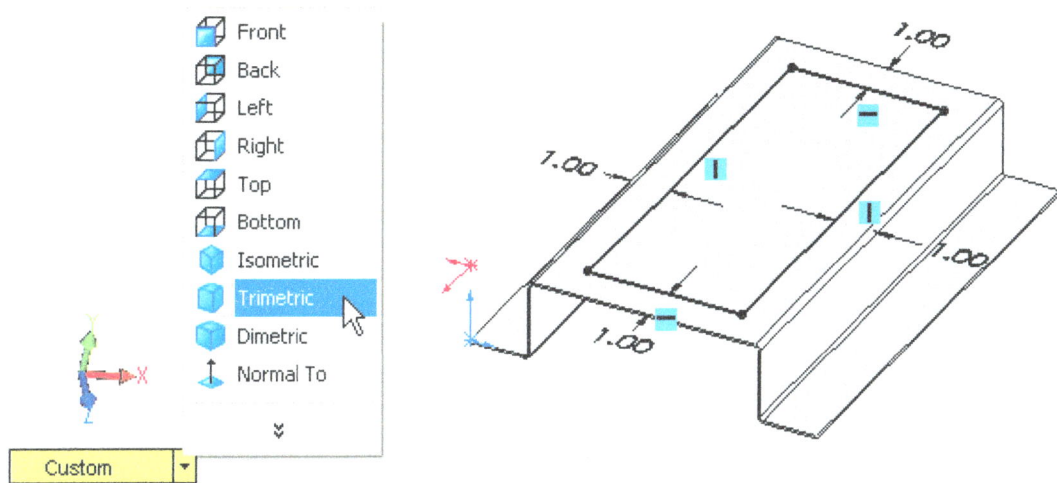

Select the **Exit Sketch** icon in the CommandManager or in the upper right corner of the graphics area.

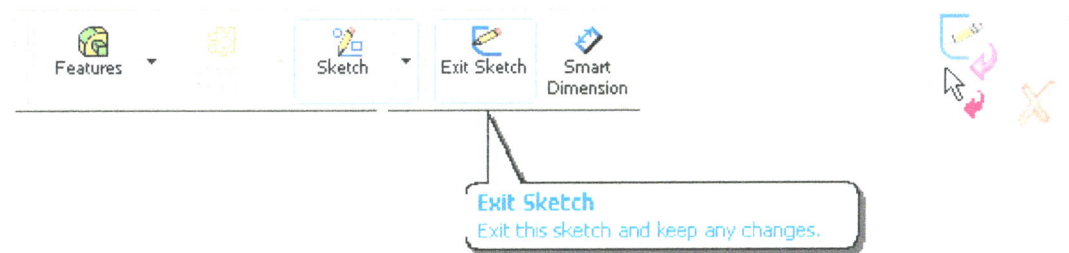

In the **Cut-Extrude** PropertyManager check the **Link to thickness** check box.

Select the green check mark button to accept the settings and create the Cut-Extrude.

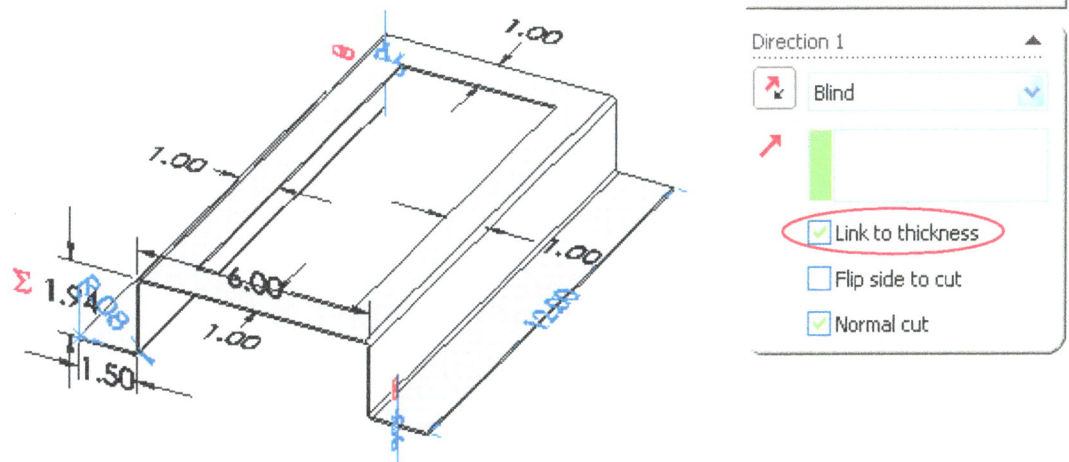

Adding Holes to the Part

In adding holes to the Top Hat, you want to try to dimension them so that the holes will be correct even if the other part dimensions are changed. To create a hole select the **Simple Hole** icon from the "Sheet Metal" toolbar or pull down the "Insert" menu and select **Features – Hole – Simple**.

Simple Hole
Creates a cylindrical hole on a planar face.

When prompted in the PropertyManager to select a location on a planer face for the center of the hole, pick the top of the lower right flange of the Top Hat.

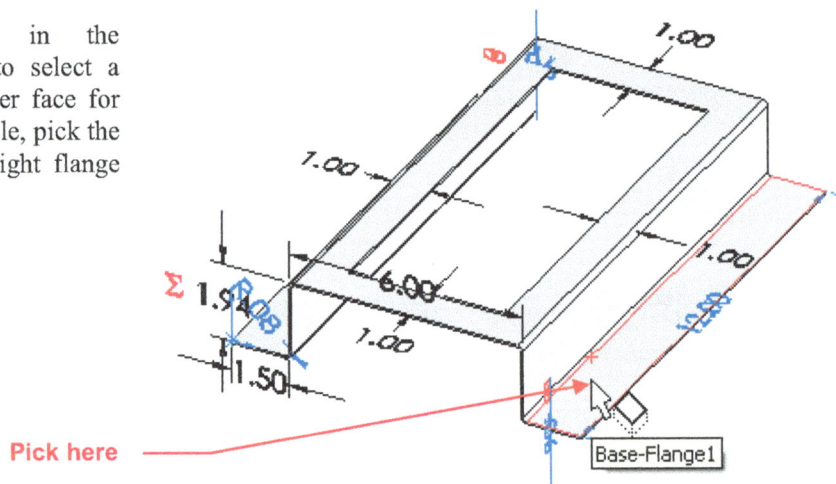

Pick here

Base-Flange1

In the **Hole** PropertyManager, enter '**0.50**' for the diameter.

Check the **Link to thickness** check box. This option ensures that the hole is through the thickness of the sheet metal, no matter what gauge or thickness is specified.

Select the green check mark button to accept the settings and create the hole.

Since the hole is placed at the cursor location, you need to dimension the sketch for more accuracy.

In the FeatureManager design tree, right click on **Hole1** and select **Edit Sketch**.

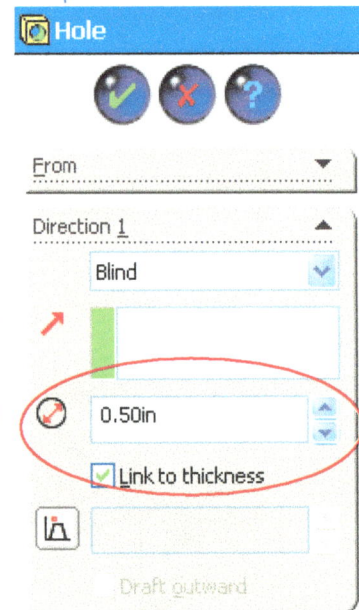

Select the **Smart Dimension** icon in the CommandManager, or pull down the "Tools" menu and select **Dimensions – Smart**.

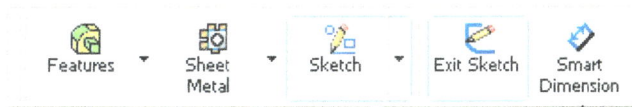

The cursor changes from ⬚ to ⬚.

Select the center point of the circle (a large red dot appears) and the corner of the part (a red dot appears). Move the cursor below the part and click to create the dimension. Make sure that the desired type of dimension is shown before placing the dimension. Enter the value '**.75**' for this dimension.

Next, add another dimension from the center point of the circle and the corner of the part as shown to the right.

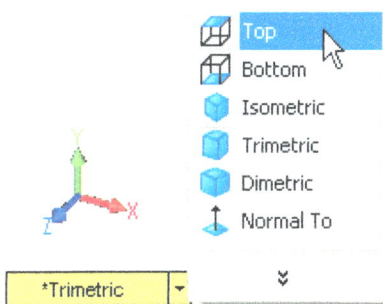

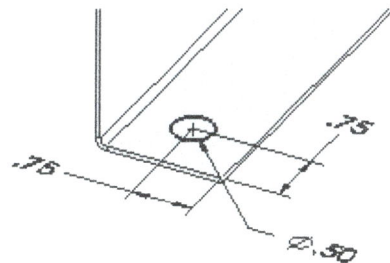

In the bottom left corner of your graphics area, change the View orientation by selecting the pull down arrow and selecting **Top**.

To add holes to the other corners, just create more circles in the sketch.

To do this, select the **Circle** icon from the "Sketch" toolbar, or pull down the "Tools" menu and select **Sketch Entities – Circle**.

Make sure that **Center creation** is selected in the **Circle** PropertyManager.

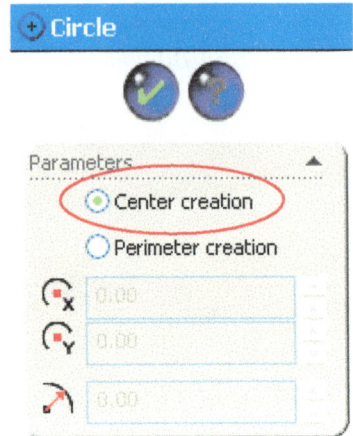

Create three circles using the **Circle** sketch command.

To do this, simply place the cursor at the location where you want the center of the circle. Click once to set this location then move the cursor. A preview of the circle as well as its radius value is shown as the cursor moves. Click a second time to create the circle. Again, the specific size and location does not matter since dimensions and relations will be added to the circles to achieve the desired locations and sizes.

Dashed blue lines appear when the cursor is aligned with an existing item.

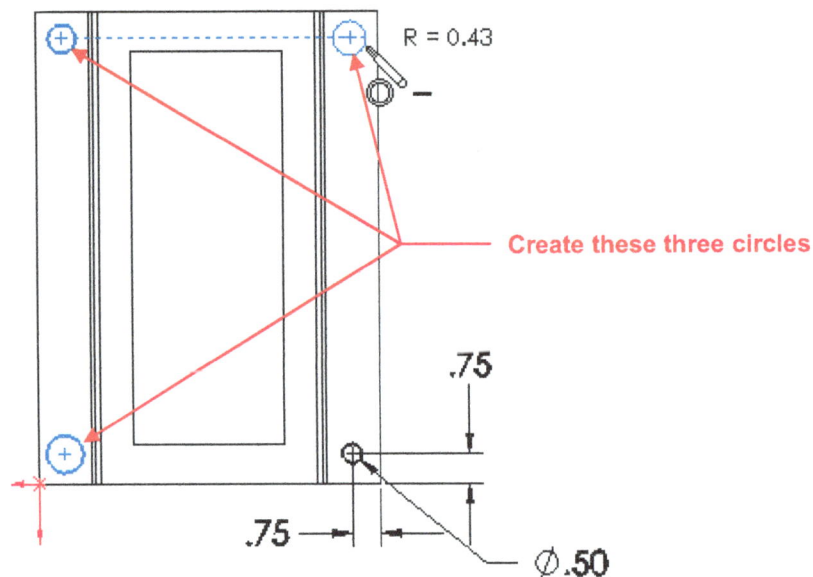

R = 0.43

Create these three circles

.75

.75

⌀.50

Select the **Add Relation** icon from the "Sketch" toolbar, or pull down the "Tools" menu and select **Relations – Add**.

Circle   Centerpoint Arc   Tangent Arc   3 Point Arc   Sketch Fillet   Centerline   Spline   Point   Add Relation

Add Relation
Controls the size or position of entities with constraints such as concentric or vertical.

Select the four circles. The circles turn green when they are selected. The last circle created may already be selected for you. Then, select the **Equal** button in the **Add Relations** PropertyManager.

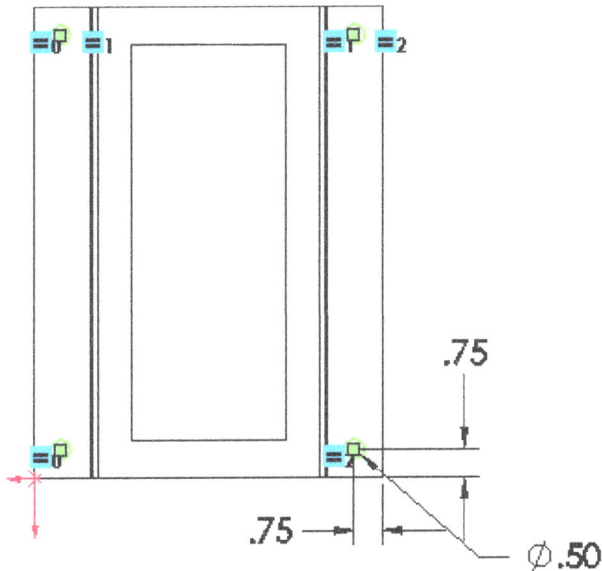

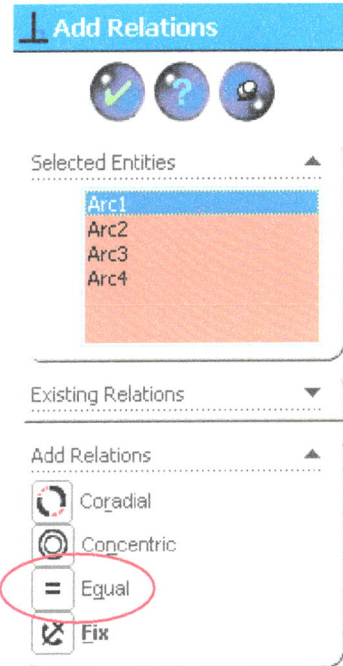

Next, select the center points of the two circles on the right hand side of the part and select the **Vertical** button. Remember, the center points will display a large red dot when the cursor is on them. The dot turns green when it is selected.

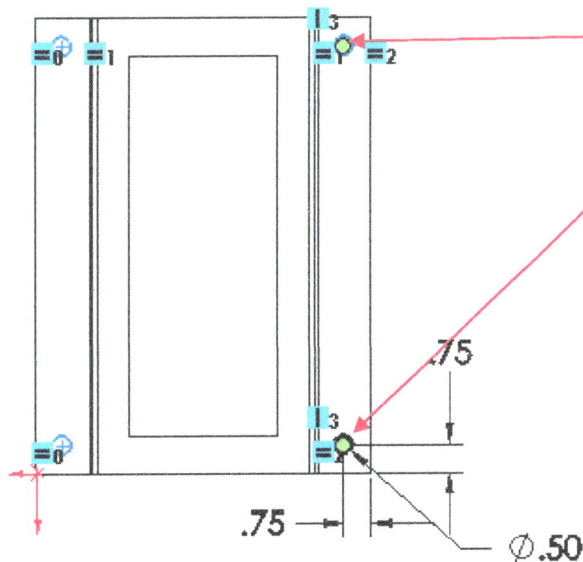

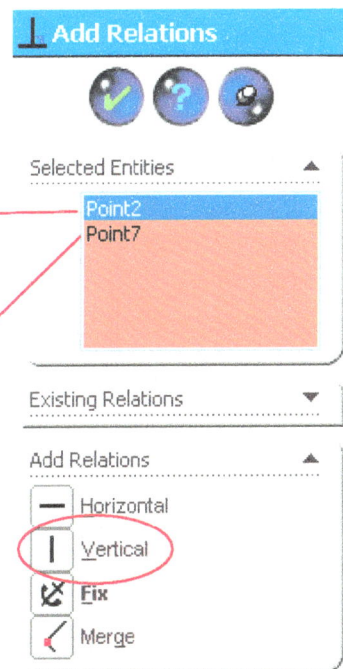

To delete any **Selected entities** from the list, you may either select the item again or you may highlight it in the **Add Relations** PropertyManager and press the **Delete** key on the keyboard. The geometry is not deleted. It is just removed from the selected entities list.

Select the center point of the lower left circle followed by the center point of the lower right circle, and then select the **Horizontal** button. Take a look at the **Selected Entities** list in the **Add Relations** PropertyManager dialog. There should be two points listed as shown on the right.

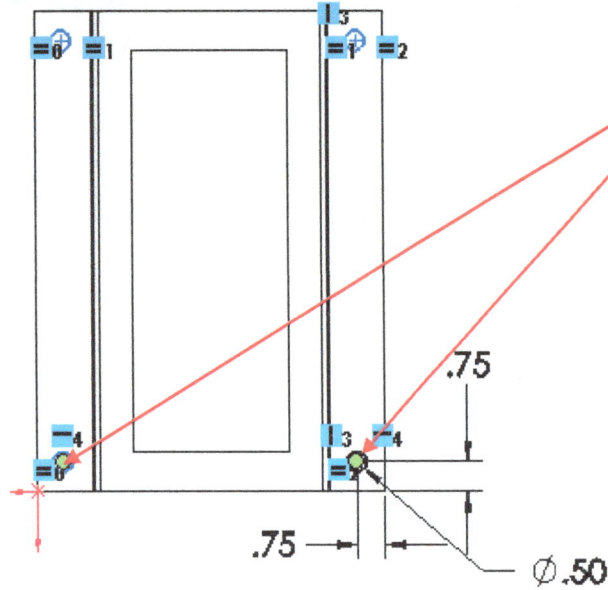

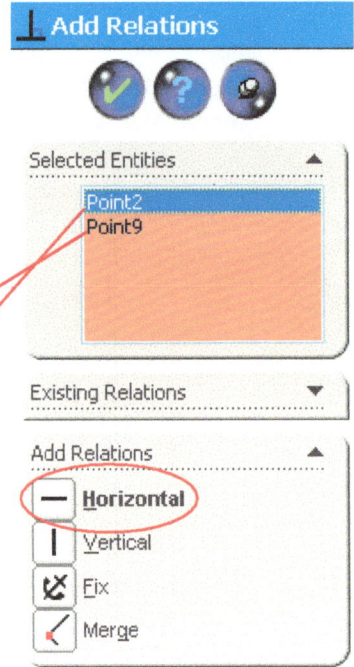

Next, select the center point of the top left circle and the center point of the bottom left circle and select the **Vertical** button.

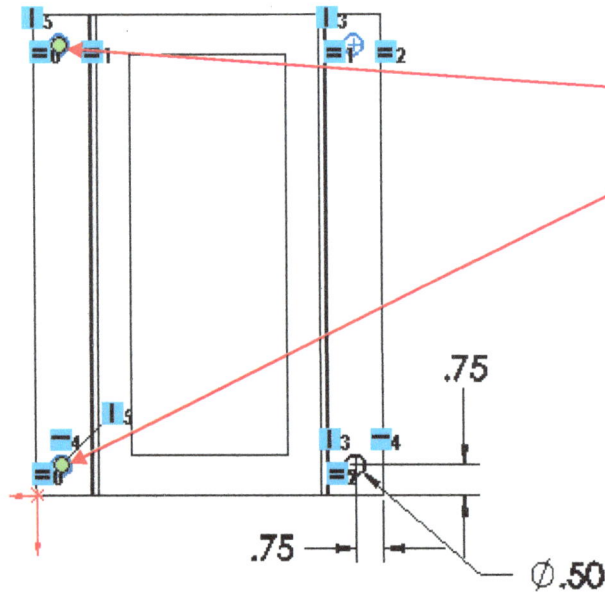

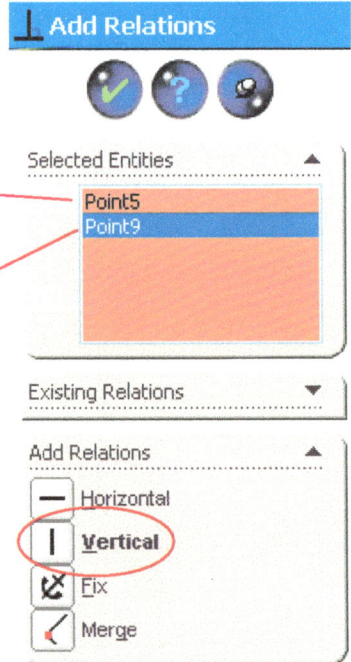

Finally, select the center point of the top right circle and the center point of the top left circle and select the **Horizontal** button.

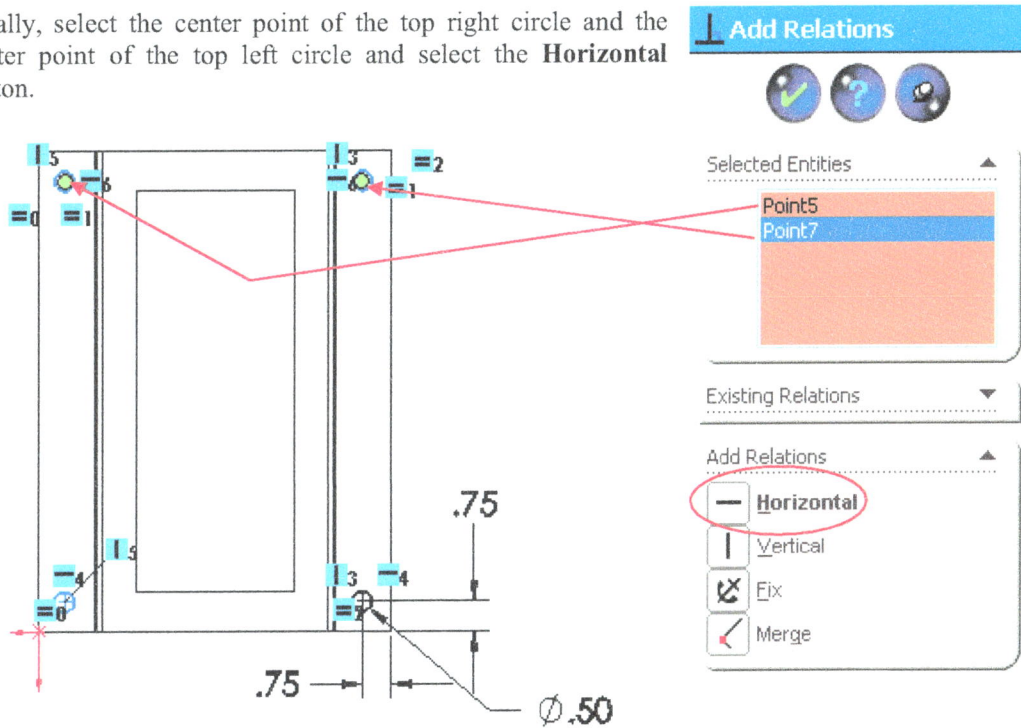

With all the relations added, select the **Smart Dimension** icon in the CommandManager, or pull down the "Tools" menu and select **Dimensions – Smart**.

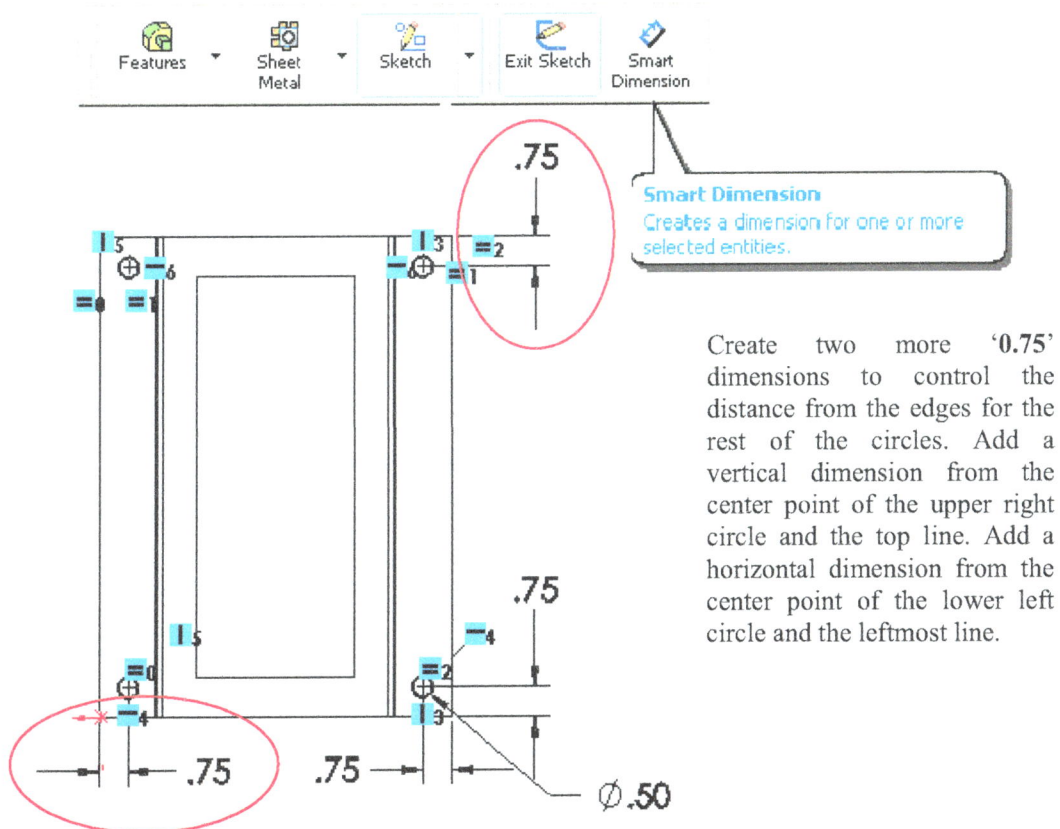

Create two more '**0.75**' dimensions to control the distance from the edges for the rest of the circles. Add a vertical dimension from the center point of the upper right circle and the top line. Add a horizontal dimension from the center point of the lower left circle and the leftmost line.

Select the **Exit Sketch** icon in the CommandManager or in the upper right corner of the graphics area.

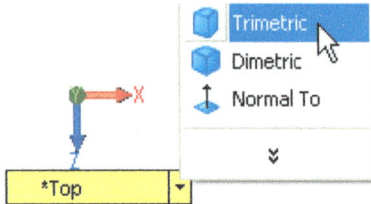

In the bottom left corner of your graphics area, change the View orientation by selecting the pull down arrow and selecting **Trimetric**.

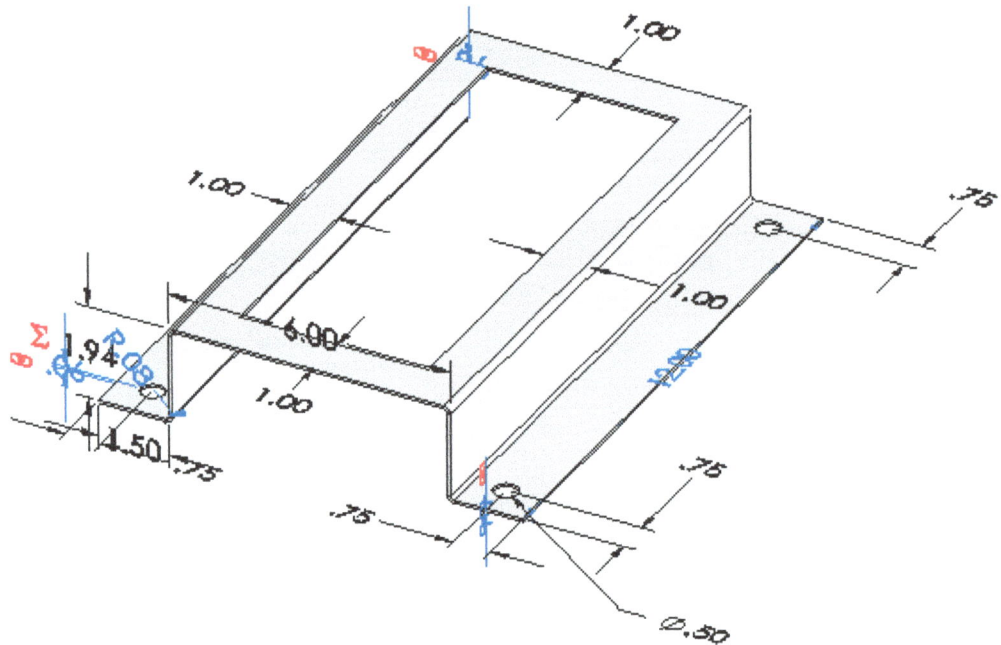

Save the Part

Select the **Save** icon on the "Standard" toolbar, or select **Save** from the "File" pull down menu.

The **Save As** dialog box appears.

In the **File name** box, type the name of the drawing number. For this chapter, use '**49426-231**' and select **Save**.

The extension **.sldprt** is added to the filename, and the file is saved. File names are not case sensitive. That is, files named **TOP HAT.sldprt**, **Top Hat.sldprt**, and **top hat.sldprt** are all the same file.

Changing the Part

You may change any of the part dimensions; material thickness, bend radius, width, or length, by double clicking on the appropriate dimension and changing the value.

Select the **Rebuild** icon in the "Standard" toolbar to see the changes, or press the keyboard shortcut **Ctrl+B**.

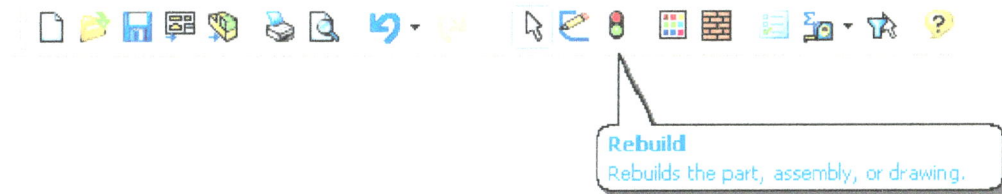

Rebuild
Rebuilds the part, assembly, or drawing.

Since the height is now part of an equation, when you double click on the dimension, the **Modify** dialog box is a little different. Simply select **Edit Equation** from the pull down menu.

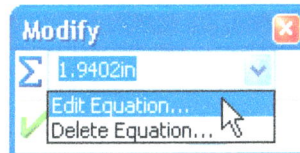

Modify
Σ 1.9402in
Edit Equation...
Delete Equation...

Change the height value in the equation to a new value, like '**3**,' in the **Edit Equation** dialog box.

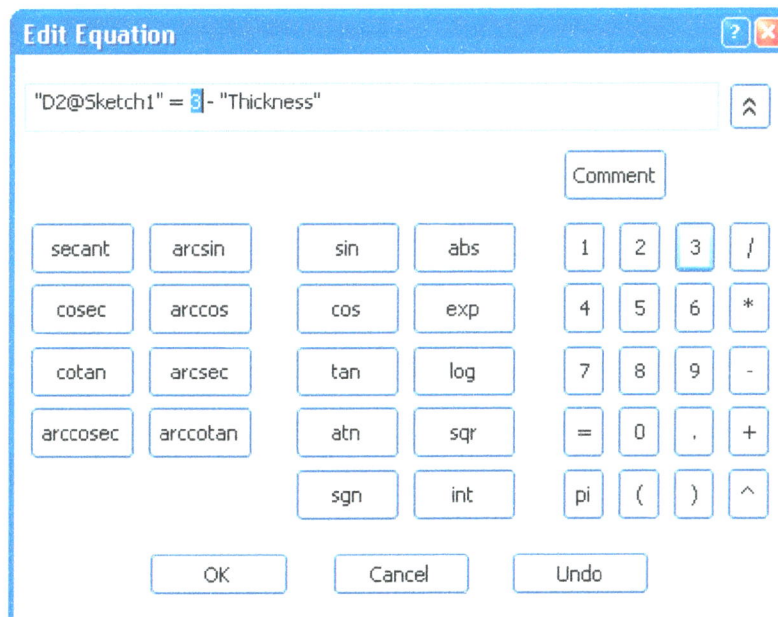

Edit Equation

"D2@Sketch1" = 3 - "Thickness"

| Comment |

secant	arcsin		sin	abs		1	2	3	/
cosec	arccos		cos	exp		4	5	6	*
cotan	arcsec		tan	log		7	8	9	-
arccosec	arccotan		atn	sqr		=	0	,	+
			sgn	int		pi	(	)	^

| OK | | Cancel | | Undo |

Select the **OK** button in the **Edit Equation** dialog box and select the **OK** button in the **Equations** dialog box.

Select the **Rebuild** icon in the "Standard" toolbar to see the change, or press the keyboard shortcut **Ctrl+B**.

Rebuild
Rebuilds the part, assembly, or drawing.

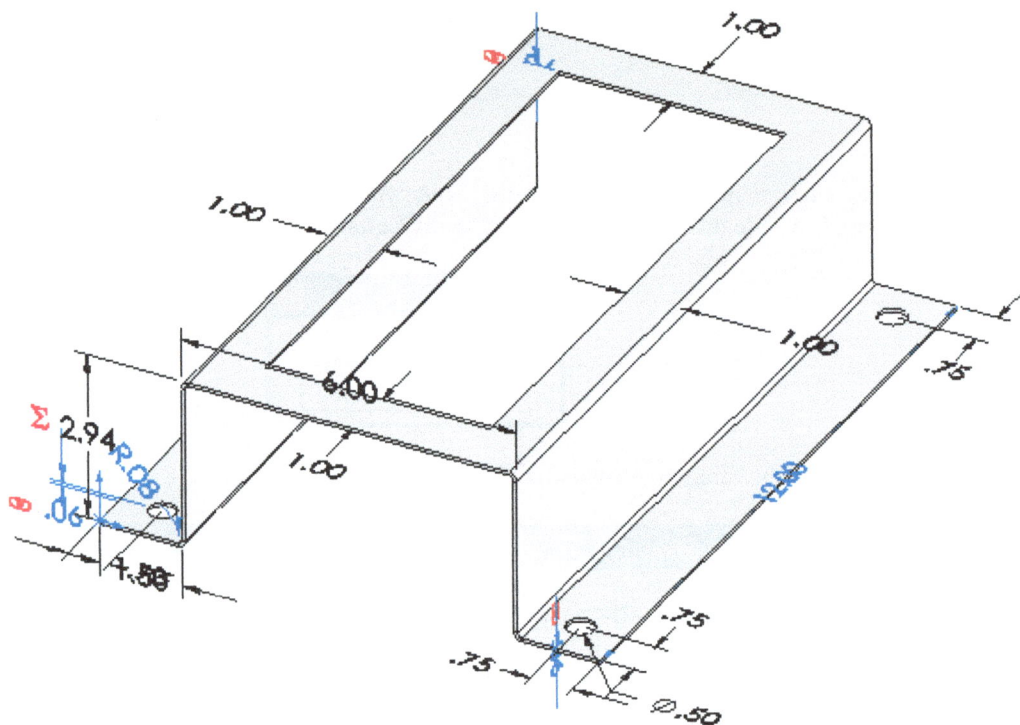

Closing the File

Select **Close** from the "File" pull down menu.

Select **Yes** when prompted to **Save changes to 49426-231.SLDPRT?** to save your changes.

Select **No** when prompted to **Save changes to 49426-231.SLDPRT?** to not save your changes.

Chapter 2

Four Sided Box

A common part is the four sided box. This part has numerous variations, with open corners, closed corners, and overlapping corners. Some variations have additional flanges and tabs. But they all start with a basic box.

This chapter creates a basic box with open corners and then shows you how to overlap the side and end flanges to close the corner. An equation is added to control the gap in the corners and to ensure the flat pattern is squared up for punching.

The part is created using outside to outside dimensions. However inside dimensions are just as easy to create. At the end of the chapter, it shows how to change the dimensions from outside to inside.

Create a New Part Document

To begin this chapter, open a new part document.

Select the **New** icon in the "Standard" toolbar, or pull down the "File" menu and select **New**.

New
Creates a new document.

The **New SolidWorks Document** dialog box appears.

Select **Part** and then select **OK**. A new part window appears.

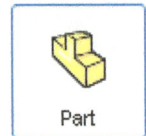

Part

Create the Base Flange

The first feature that you will create in this part is a base flange created from a sketched rectangular profile. Begin by sketching a rectangle.

To do this, select the **Sheet Metal** icon in the control area of the CommandManager. Then, select the **Base-Flange/Tab** icon from the toolbar, or pull down the "Insert" menu and select **Sheet Metal – Base Flange**.

Features   Sheet Metal   Sketch   Base-Fla...   Edge Flange   Miter Flange

Base-Flange/Tab
Creates a sheet metal part or adds material to an existing sheet metal part.

The **Front**, **Top**, and **Right** planes appear,

and the cursor changes to _____ .

You are prompted in the PropertyManager, on the left hand side of the screen, to select a plane on which to sketch the part cross-section. As you move the cursor over a plane, the border of the plane is highlighted.

Move the cursor so that the **Top Plane** is highlighted. Click the left mouse button to select the **Top Plane**.

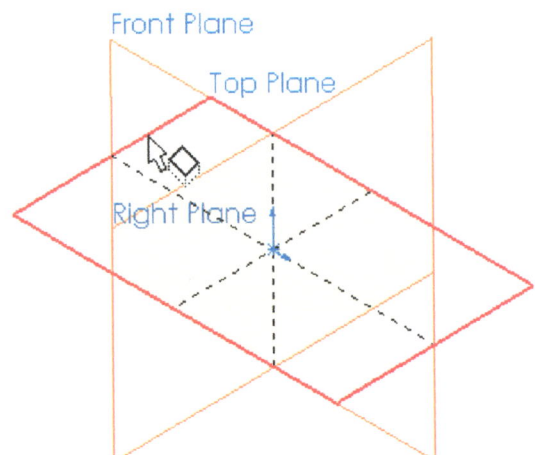

Front Plane

Top Plane

Right Plane

The display changes so that the **Top Plane** is facing you and a new sketch is opened on this plane. The CommandManager dynamically updates to show the "Sketch" toolbar.

Select the **Rectangle** icon in the CommandManager, or pull down the "Tools" menu and select **Sketch Entities – Rectangle**.

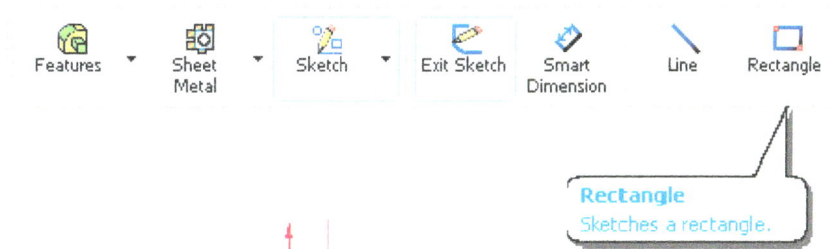

Move the cursor to the origin.

A large red dot appears when you are there.

The cursor is on the origin when the cursor changes from   to

Pick the origin by clicking the left mouse button.

Move the cursor up and to the right to create a rectangle. As you move the cursor, an outline of the rectangle is displayed.

To complete the rectangle, click the left mouse button. The size of the rectangle does not matter. You will use the **Smart Dimension** tool to dimension the sketch to the desired size.

Adding Dimensions

Select the **Smart Dimension** icon in the CommandManager, or pull down the "Tools" menu and select **Dimensions – Smart**.

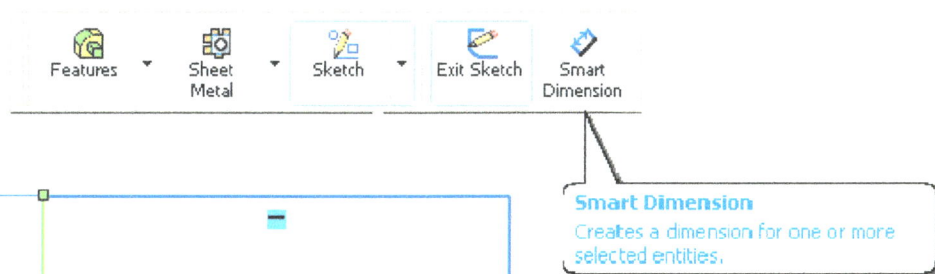

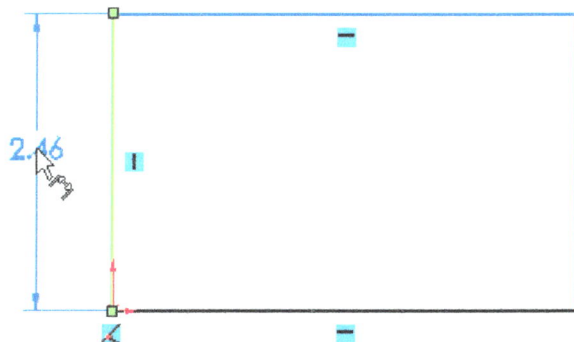

Select the left vertical line. Then, move the cursor to the left to where you want to place the dimension and click the left mouse button.

The **Modify** dialog box appears and the dimension value is highlighted.

Type the number '**6**' and then select ✔ or press the **Enter** key on the keyboard.

Select the bottom horizontal line.

Then move the cursor down and click the left mouse button to place the dimension.

The **Modify** dialog box appears again.

Type the number '**8**' and then select ✔ or press the **Enter** key on the keyboard.

Select the **Zoom to Fit** icon in the "View" toolbar, or pull down the "View" menu and select **Modify – Zoom to Fit** so you can see the entire rectangle and center it in the graphics area.

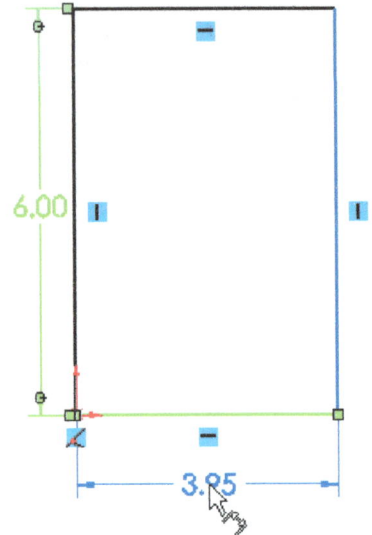

Zoom to Fit
Zooms the model to fit the window.

All four lines should now be black. The status bar in the lower-right corner of the window indicates that the sketch is now fully defined.

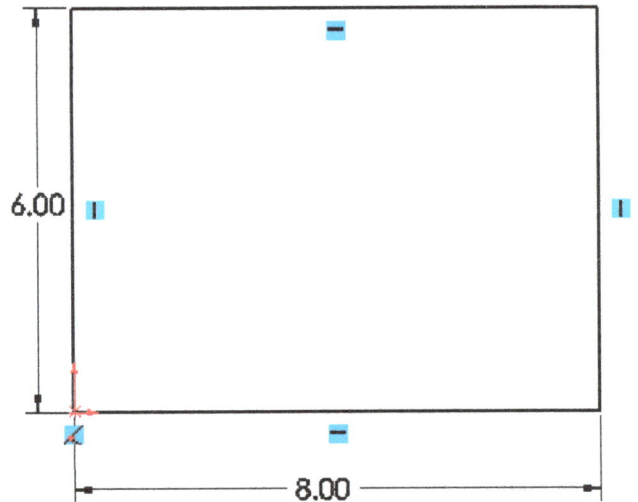

Select the **Exit Sketch** icon in the CommandManager or in the upper right corner of the graphics area.

Exit Sketch
Exit this sketch and keep any changes.

Completing the Base Flange

The **Base Flange** PropertyManager appears, the graphics area changes to the standard **Trimetric** view, and a preview of the base flange appears.

In the **Base Flange** PropertyManager under **Sheet Metal Parameters**, set the **Thickness** to '**0.048in**'.

Make certain the **Reverse direction** check box is checked to make sure that the preview of the extrusion is in the Y-positive direction (The yellow rectangle will be above the green rectangle).

Select the green check mark button at the top of the **Base Flange** PropertyManager to create the part.

In the FeatureManager design tree, right click on the **Sheet-Metal1** item and select **Edit Feature** from the menu.

In the **Sheet-Metal1** PropertyManager, enter '**0.05in**' for the default **Bend Radius**.

Select the green check mark button at the top of the **Sheet-Metal1** PropertyManager.

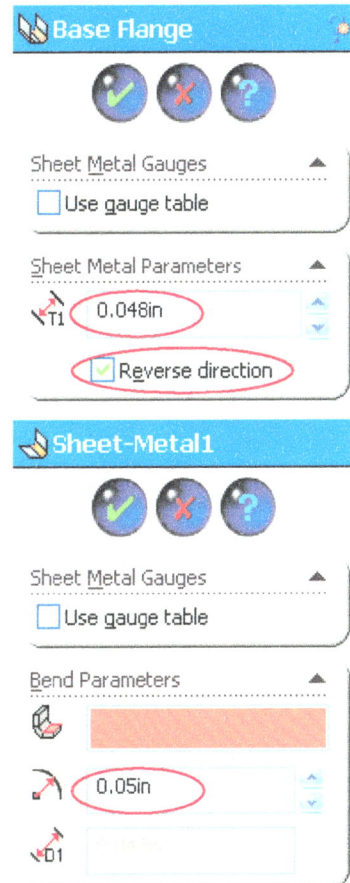

Create the Sides of the Box

Select the **Edge Flange** icon in the CommandManager, or pull down the "Insert" menu and select **Sheet Metal – Edge Flange**.

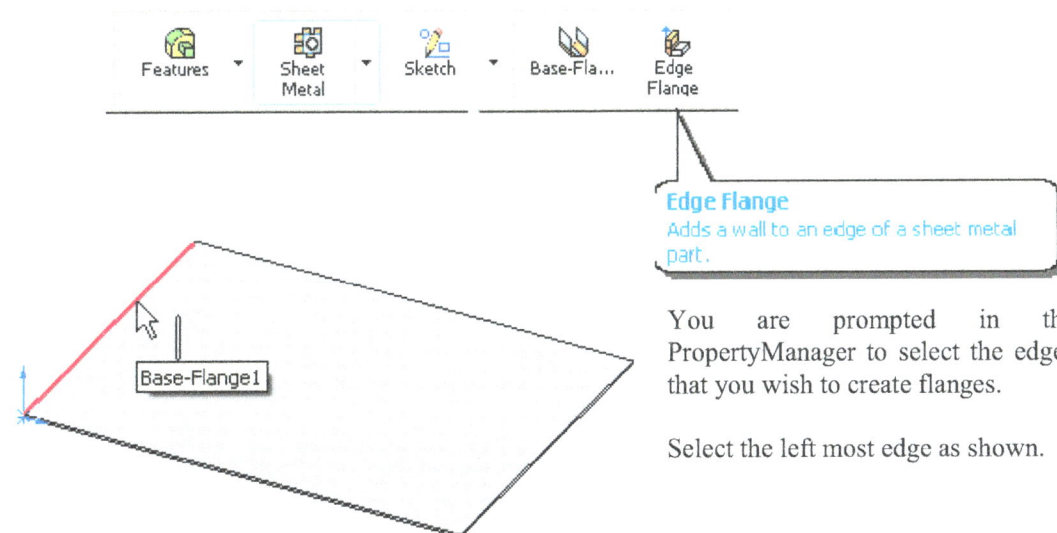

Edge Flange
Adds a wall to an edge of a sheet metal part.

You are prompted in the PropertyManager to select the edges that you wish to create flanges.

Select the left most edge as shown.

A preview of the flange will be shown. Move the cursor up and click the left mouse button to set the direction of the flange to be up instead of down.

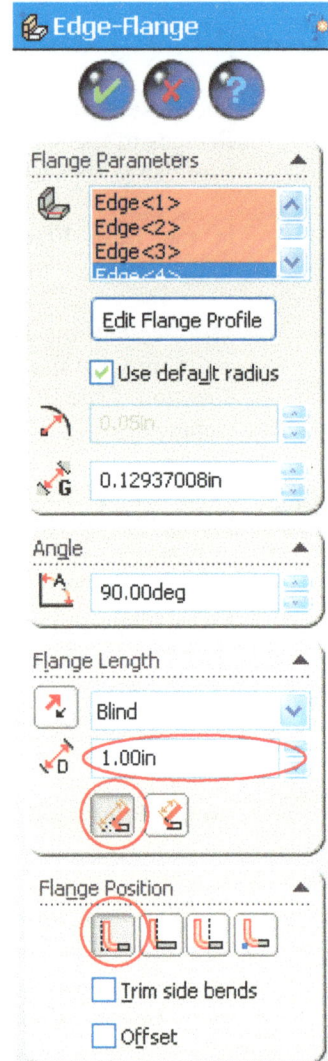

Edge-Flange

Flange Parameters

Edge<1>
Edge<2>
Edge<3>
Edge<4>

Edit Flange Profile

☑ Use default radius

0.05in

G 0.12937008in

Angle

A 90.00deg

Flange Length

Blind

D 1.00in

Flange Position

☐ Trim side bends
☐ Offset

With the direction set, you are able to select multiple edge lines to create multiple flanges. Select the remaining three edge lines of the rectangle.

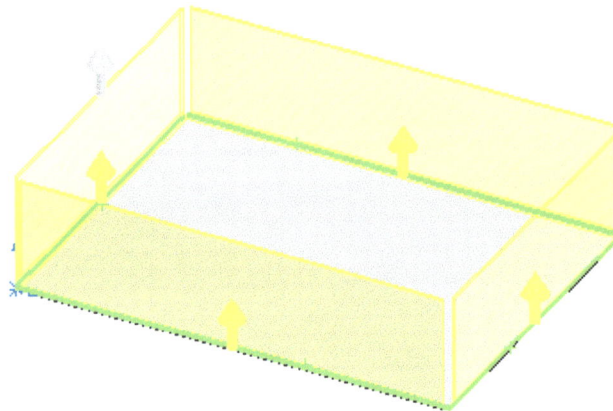

In the **Edge-Flange** PropertyManager, set the **Flange Length** to a **Blind Length** of '**1.00in**'.

Make sure that the **Outer Virtual Sharp** button is depressed.

Set the **Flange Position** to **Material Inside**.

Select the green check mark button at the top of the **Edge-Flange** PropertyManager to accept the settings and create the flanges.

If the part is off the screen, select the **Zoom to Fit** icon in the "View" toolbar, or pull down the "View" menu and select **Modify – Zoom to Fit** so you can see the entire part and center it in the graphics area.

Zoom to Fit
Zooms the model to fit the window.

Display the Dimensions

In the FeatureManager design tree, right click on **Annotations** and select **Show Feature Dimensions**. You may drag and drop dimensions with the mouse to reposition them.

Adding Equations

To make sure that the corner relief is correctly drawn, an equation must be added to the **Gap distance** dimension. The gap distance is measured diagonally, similar to an isosceles right triangle. Math formulas can be used to allow you to specify it as a horizontal/vertical distance. In this case you will set it to be equal to the bend radius.

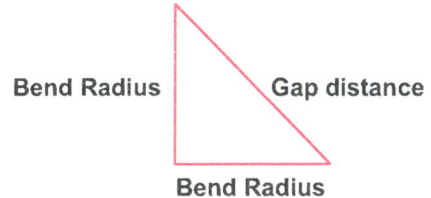

Bend Radius   **Gap distance**

Bend Radius

Gap distance = Bend Radius * √ 2

Looking at the top of the part, you can see that the ends of the side flanges are not in line with the edges of the base flange.

To fix this problem, find the **Gap distance** dimension and the **Bend Radius** dimension. Remember that you can drag and drop them to make it easier to see them.

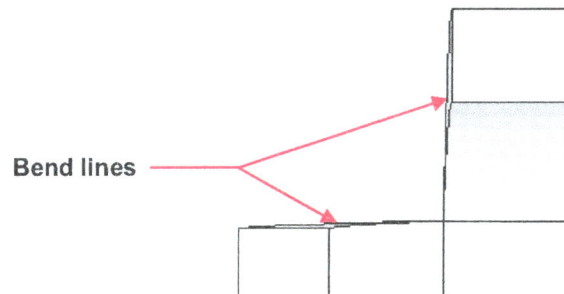

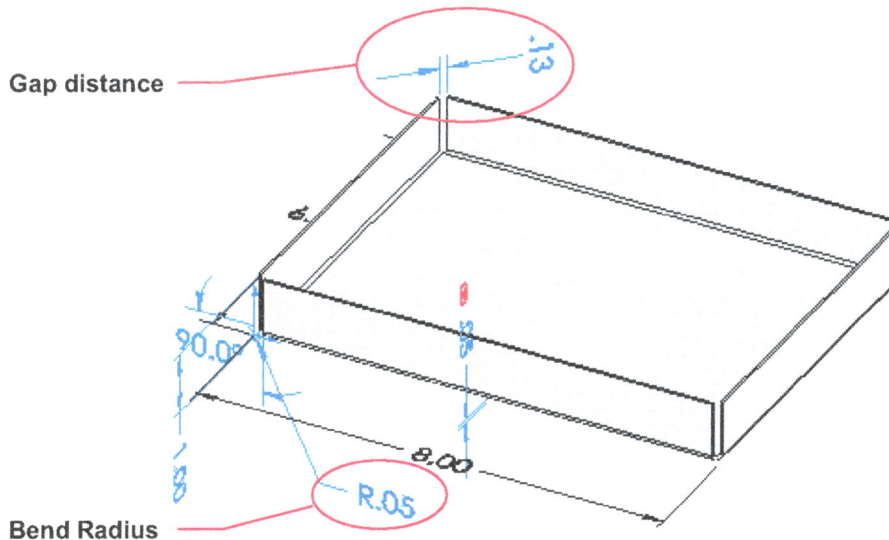

Bend lines

Gap distance

Bend Radius

After you have located the dimensions and they are in view so you will be able to select them, double click on the **Gap distance** dimension on the part.

In the **Modify** dialog box, select **Add Equation** from the pull down menu.

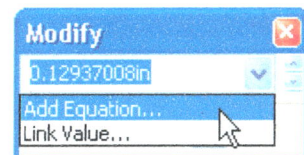

Modify
0.12937008in
Add Equation...
Link Value...

Select the **Bend Radius** dimension. You may need to drag the dialog boxes out of the way in order to see the dimension. Then, select the multiplication (*) button. Finally, select the square root (**sqr**) button followed by the **2** button. Once the equation is complete, select the **OK** button.

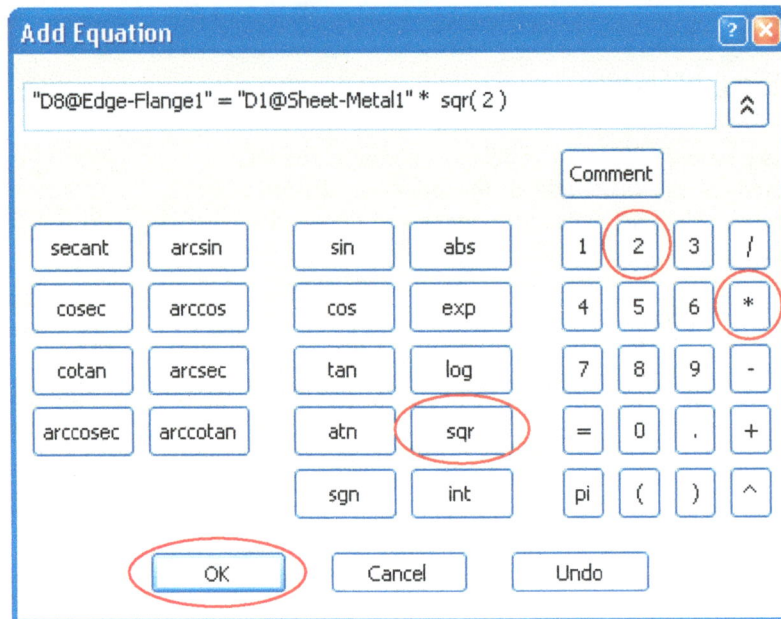

Select the **OK** button in the **Equations** dialog box.

The **Gap distance** dimension is updated. To get the part to update with the new gap distance, select the **Rebuild** icon in the "Standard" toolbar.

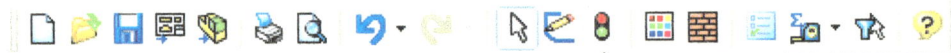

Rebuild
Rebuilds the part, assembly, or drawing.

Now, if the **Bend Radius** of the part is ever changed, the equation ensures that the **Gap distance** will always be updated to the correct dimension.

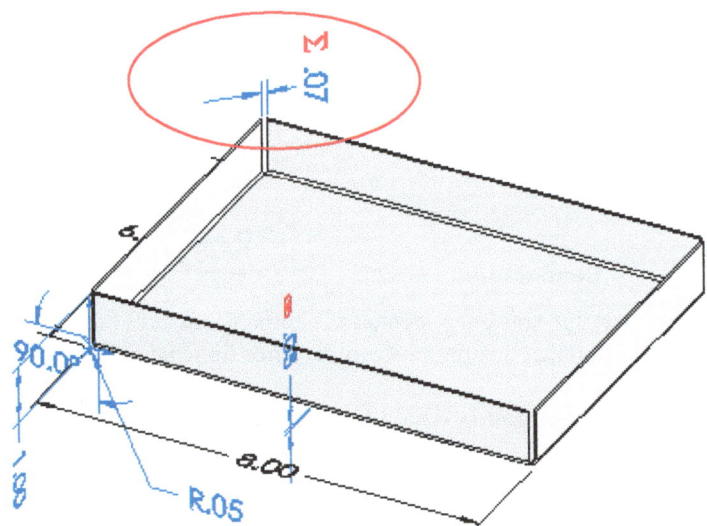

Close the Corners

To close the corners of the part, select the **Sheet Metal** icon in the control area of the CommandManager. Then, select the **Closed Corner** icon from the toolbar, or pull down the "Insert" menu and select **Sheet Metal – Closed Corner**.

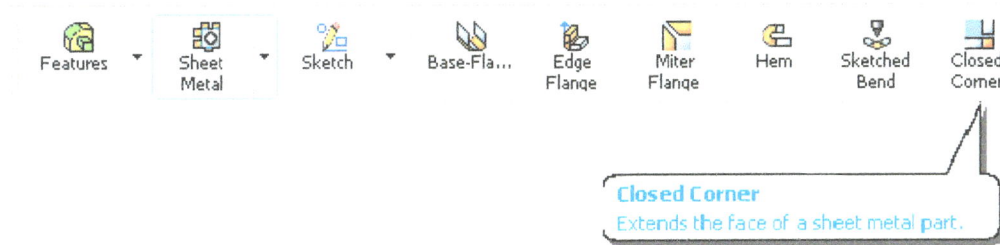

Select the **Zoom to Area** icon in the "View" toolbar, or pull down the "View" menu and select **Modify – Zoom to Area**.

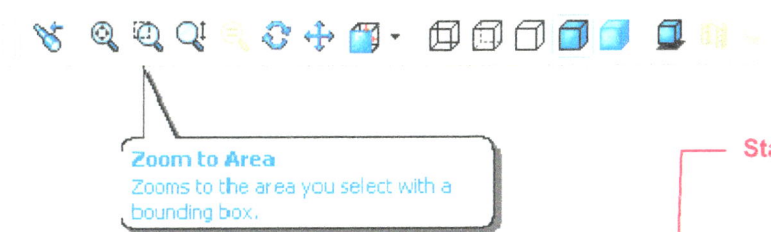

Place the cursor above and to the left of the bottom corner of the part. Press the left mouse button and drag the cursor down and to the right around the area that you want to zoom in to. When you release the mouse button the view is zoomed in to the selected area.

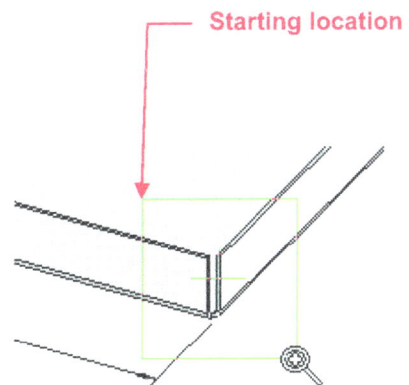

Select the **Zoom to Area** icon in the "View" toolbar again to deselect the icon, or pull down the "View" menu and select **Modify – Zoom to Area**. This returns you to the **Closed Corner** command.

You must now select the edge (face) of the side flanges that you want to extend to close the corner. For the **Faces to Extend**, select the edge face of the long, front flange as shown.

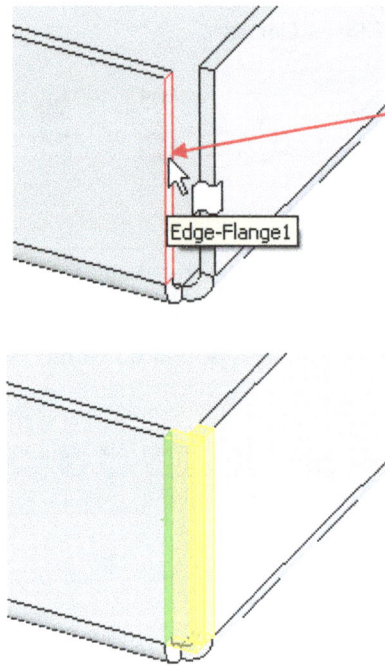

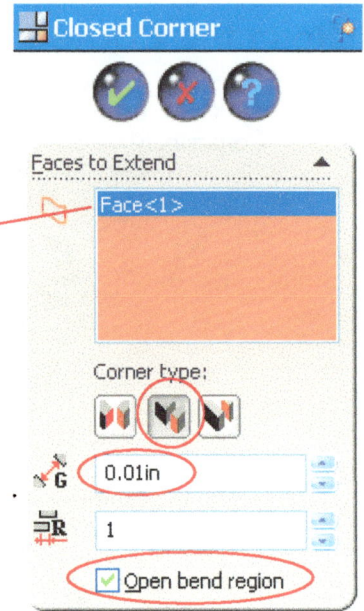

Under **Corner type**, select the **Overlap** button.

Set the **Gap distance** to '**0.01in**'.

Check the **Open bend region** check box.

Select the **Zoom to Fit** icon in the "View" toolbar, or pull down the "View" menu and select **Modify – Zoom to Fit** so you can see the entire part and center it in the graphics area.

Zoom to Fit
Zooms the model to fit the window.

SolidWorks for the Sheet Metal Guy

Zoom in and repeat the above steps for the three remaining corners. Make sure that you select the sides of the longer, 8" flanges.

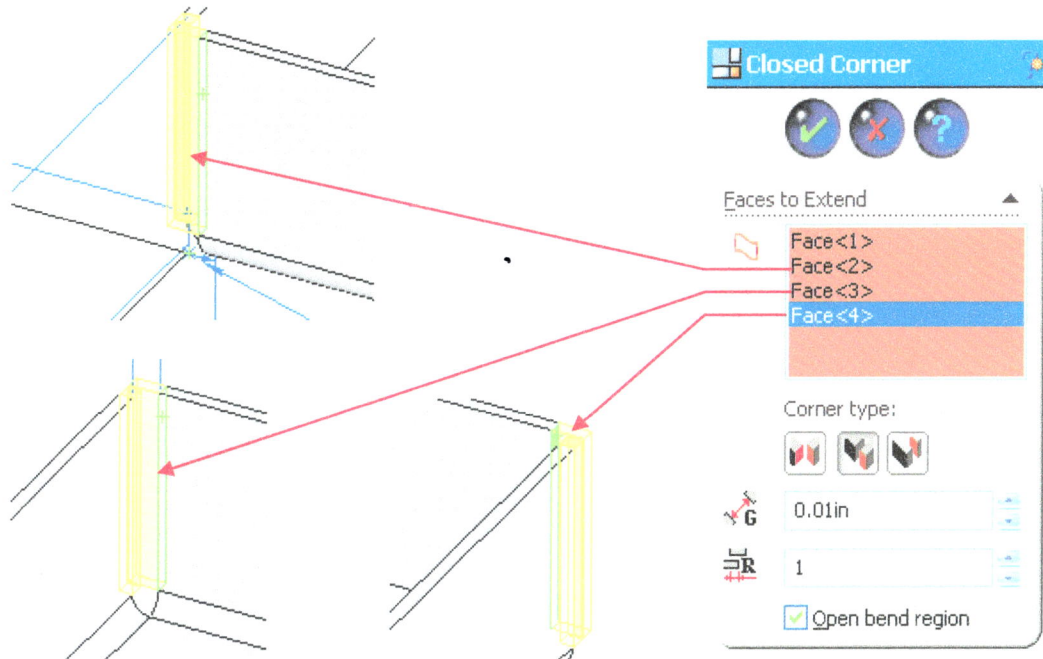

If you are having trouble selecting the correct face, right click on a face and select **Select Other**.

A list is shown of available faces (sides). Highlight the hidden face with the cursor and left-click.

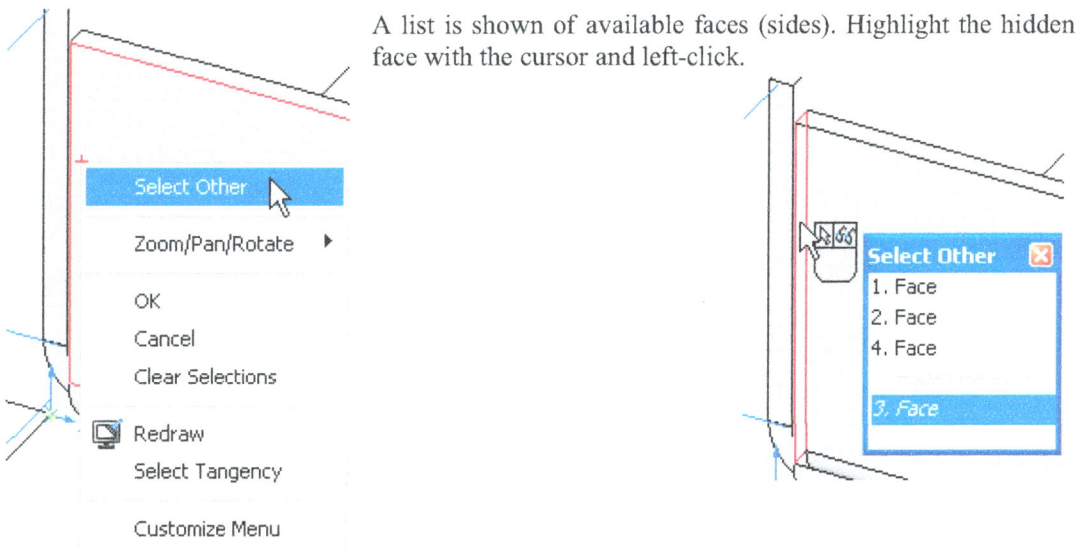

When you have selected all four faces, select the green check mark button at the top of the **Closed Corner** PropertyManager to accept the settings and close the corners.

Create the Flat Pattern

Select the **Flatten** icon from the Sheet Metal CommandManager toolbar. You may also right-click on **Flat-Pattern1** FeatureManager design tree and select **Unsuppress**.

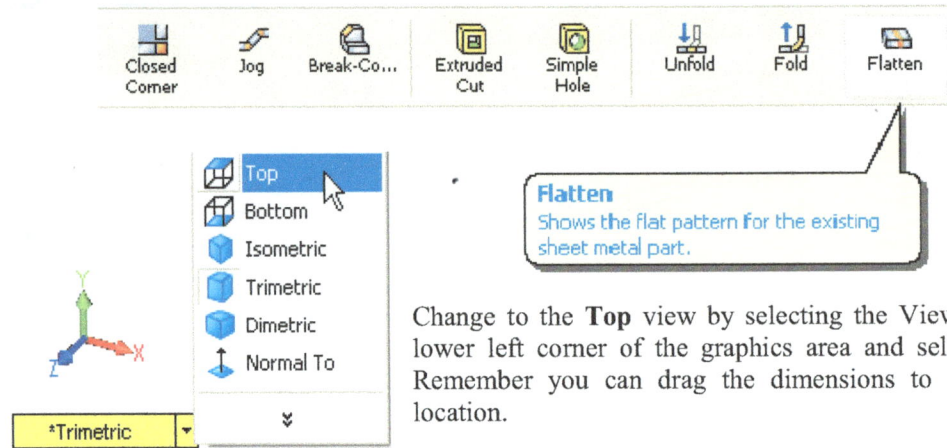

Flatten
Shows the flat pattern for the existing sheet metal part.

Change to the **Top** view by selecting the View list in the lower left corner of the graphics area and selecting **Top**. Remember you can drag the dimensions to any desired location.

When you flatten the sheet metal part, corner treatments are applied automatically by SolidWorks to create a clean, flattened sheet metal part. In this situation, we do not want the corner treatments turned on.

Using **Zoom to Area**, look at the upper right hand corner of the flat pattern.

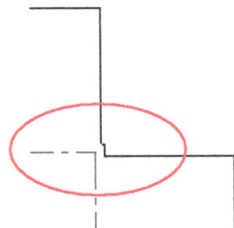

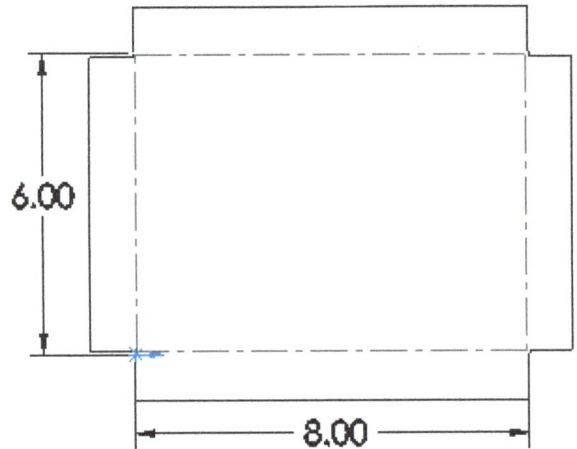

To turn off corner treatments, right-click **Flat-Pattern1**, and select **Edit Feature**.

In the **Flat-Pattern** PropertyManager, uncheck the **Corner treatment** check box.

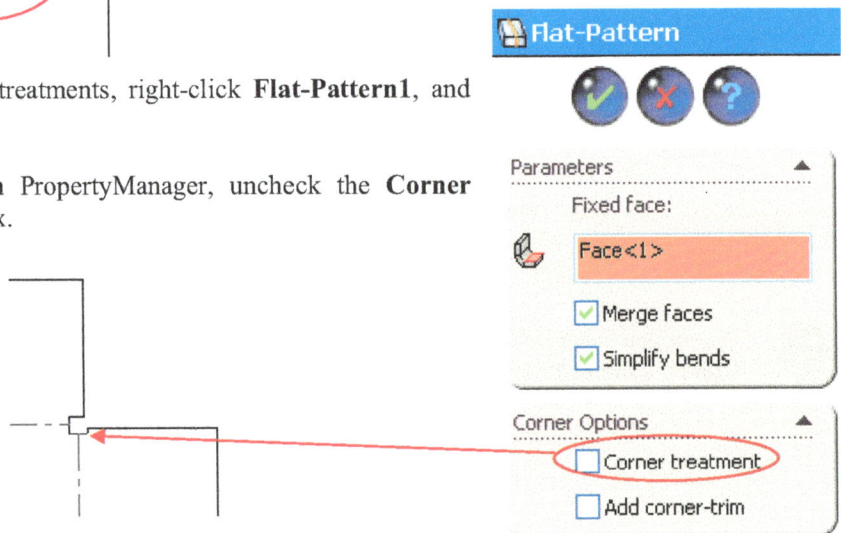

Flat-Pattern

Parameters

Fixed face:

Face<1>

☑ Merge faces

☑ Simplify bends

Corner Options

☐ Corner treatment

☐ Add corner-trim

Select the green check mark button at the top of the **Flat-Pattern** PropertyManager to accept the settings.

Select the **Flatten** icon again from the **Sheet Metal** control area of the CommandManager to fold the part back up. You may also right-click on **Flat-Pattern1** and select **Suppress**.

Flatten
Shows the flat pattern for the existing sheet metal part.

Change back to the **Trimetric** view by selecting **Trimetric** in the View list in the lower left corner of the graphics area.

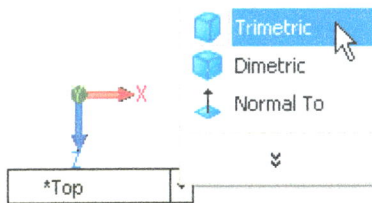

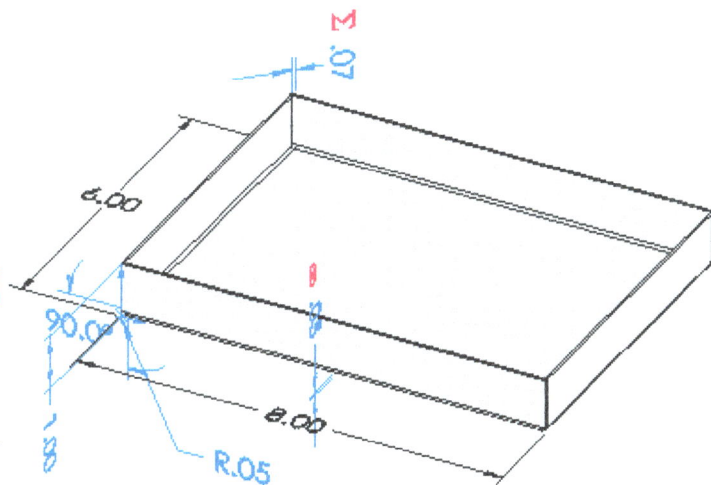

Save the Part

Select the **Save** icon in the "Standard" toolbar, or select **Save** from the "File" pull down menu.

The **Save As** dialog box appears. In the **File name** box, type the name of the drawing number. For this chapter, use '**16226-19**' and select **Save**.

Changing the Part

Again, if the material thickness, bend radius, length, width, height dimensions are changed, the key dimensions will remain constant, achieving the design intent.

Measuring Outside Versus Inside

If you need to measure the part from the inside of the material instead of from the outside, you can easily update the model.

Right-click on **Edge-Flange1** and select **Edit Feature**.

When the **Edge-Flange1** PropertyManager appears, change the **Flange Position** to **Material Outside**.

Select the green check mark button at the top of the **Edge-Flange1** PropertyManager to accept the new settings.

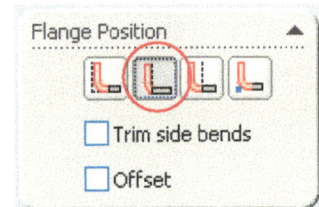

To verify that the part has updated, switch to the **Top** view by selecting the View orientation pull down arrow in the bottom left corner of your graphics area and selecting **Top**.

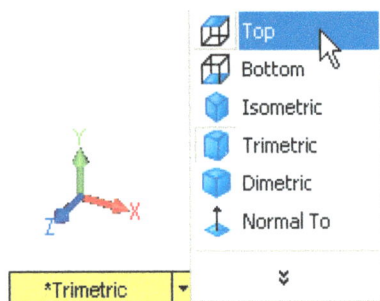

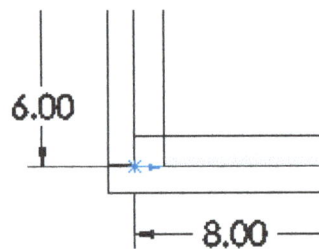

You will see that the length and width dimensions now measure to the inside surface of the flanges instead of outside to outside.

Closing the File

Select **Close** from the "File" pull down menu.

Select **Yes** when prompted to **Save changes to 16226-19.SLDPRT?** to save your changes.

Select **No** when prompted to **Save changes to 16226-19.SLDPRT?** to not save your changes.

Chapter 3

Box with Top

Moving beyond the standard box, this part has additional flanges which close the top and requires a tight fit in the corners for welding.

The Miter Flange command is ideal for this type of part. It allows you to define the profile (cross section) of the flanges and apply it to multiple edges of the part.

Finally, an equation is used to account for the material thickness in the height of the part. However, in this case, the top flange is at an angle and you must use some trigonometry to get it right.

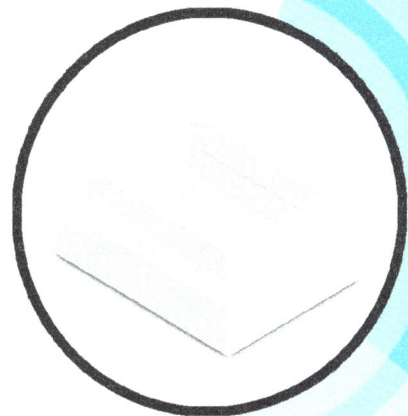

Create a New Part Document

Create a new part document.

Select the **New** icon in the "Standard" toolbar, or pull down the "File" menu and select **New**.

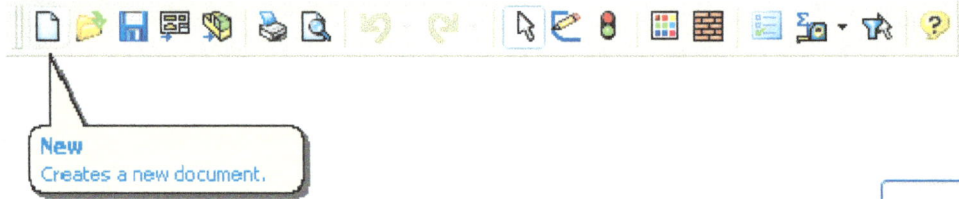

New
Creates a new document.

The **New SolidWorks Document** dialog box appears.

Select **Part** and then select **OK**. A new part window appears.

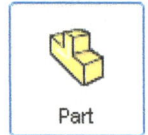

Part

Create the Base Flange

Create a base flange as we did in the previous chapter.

To do this, select the **Sheet Metal** icon in the control area of the CommandManager. Then, select the **Base-Flange/Tab** icon from the toolbar, or pull down the "Insert" menu and select **Sheet Metal – Base Flange**.

Features   Sheet Metal   Sketch   Base-Fla...   Edge Flange   Miter Flange

Base-Flange/Tab
Creates a sheet metal part or adds
material to an existing sheet metal part.

The **Front**, **Top**, and **Right** planes appear.

You are prompted in the PropertyManager to select a plane on which to sketch the feature cross-section. As you move the cursor over a plane, the border of the plane is highlighted.

Move the cursor so that the **Top Plane** is highlighted. Click the left mouse button to select the **Top Plane**.

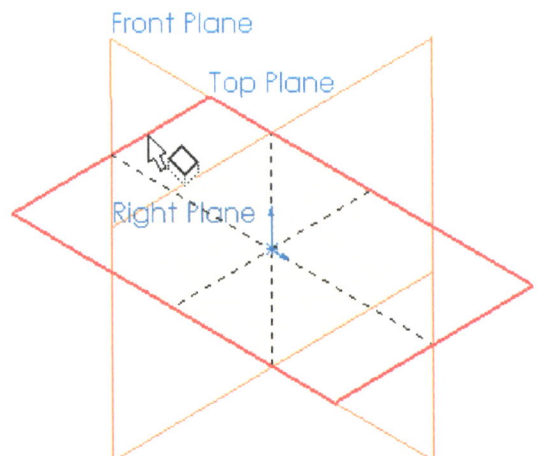

Front Plane
Top Plane
Right Plane

Select the **Rectangle** icon in the CommandManager, or pull down the "Tools" menu and select **Sketch Entities – Rectangle**.

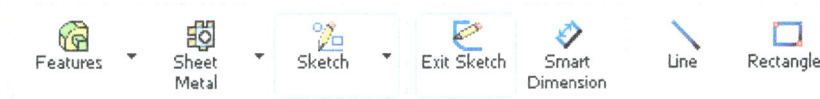

Move the cursor to the origin.

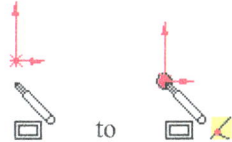

When the cursor changes from [icon] to [icon] click the left mouse button.

Move the cursor up and to the right to create a rectangle. Click the left mouse button to complete the rectangle.

Dimension the Rectangle

Select the **Smart Dimension** icon in the CommandManager, or pull down the "Tools" menu and select **Dimensions – Smart**.

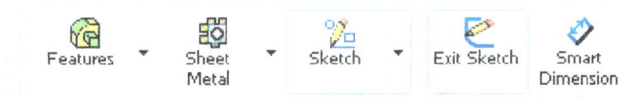

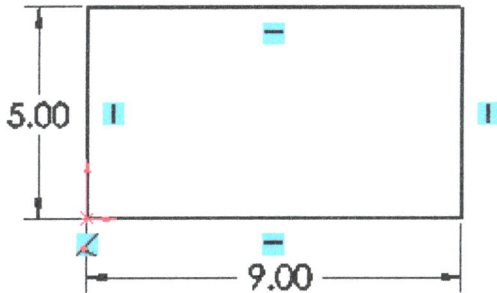

Select the left vertical line of the rectangle and place a '**5.00**' dimension.

Then, select the bottom horizontal line and place a '**9.00**' dimension.

Select the **Exit Sketch** icon in the CommandManager or in the upper right corner of the graphics area.

Finish Creating the Base Flange

The **Base Flange** PropertyManager appears, the graphics area changes to the standard **Trimetric** view, and a preview of the base flange appears.

In the **Base Flange** PropertyManager under **Sheet Metal Parameters**, set the **Thickness** to '**0.032in**'.

Check the **Reverse direction** check box to make sure that the preview of the extrusion is in the Y-positive direction.

Select the green check mark button at the top of the **Base Flange** PropertyManager.

In the FeatureManager design tree, right click on the **Sheet-Metal1** item and select **Edit Feature** from the menu.

In the **Sheet-Metal1** PropertyManager, enter '**0.03in**' for the default **Bend Radius**.

Select the green check mark button at the top of the **Sheet-Metal1** PropertyManager to finish creating the base flange.

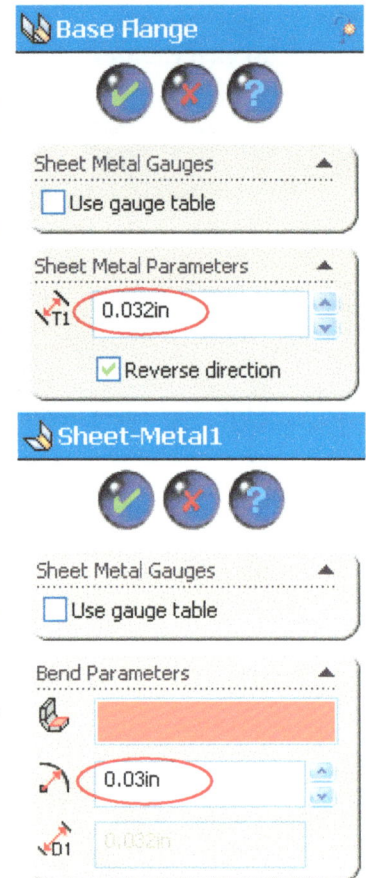

Create the Flanges

Select the **Miter Flange** icon in the CommandManager, or pull down the "Insert" menu and select **Sheet Metal – Miter Flange**.

You are prompted in the PropertyManager to select a plane, a planar face or an edge on which to sketch the feature cross-section. What you will actually be sketching is the profile of the flanges around the box.

Select the front edge face of the **Base Flange** as shown.

You may need to select the **Zoom to Area** icon in the "View" toolbar, or pull down the "View" menu and select **Modify – Zoom to Area**.

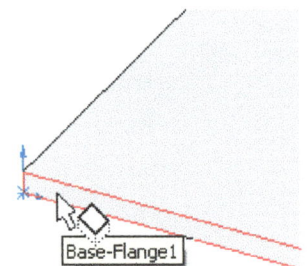

A sketch is created automatically for you. Change the display view to **Normal To** the sketch by selecting **Normal To** in the View list in the lower left corner of the graphics area.

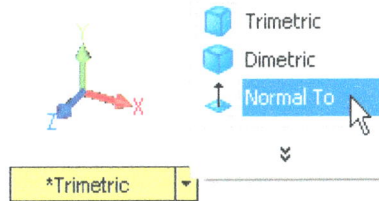

Select the **Line** icon in the CommandManager, or pull down the "Tools" menu and select **Sketch Entities – Line**.

Use the **Line** tool to create the side profile of the flange as shown.

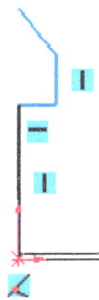

Start by creating a vertical line beginning at the origin and going up, in the positive Y direction.

Then, create a short horizontal line to the right and a short vertical line going up in the positive Y direction.

Finally, create an angled line in the negative x direction.

To help account for the material thickness when dimensioning the profile cross section, a point will be created and dimensioned along with the profile.

Select the **Point** icon in the CommandManager, or pull down the "Tools" menu and select **Sketch Entities – Point**.

Create a point above the angled line.

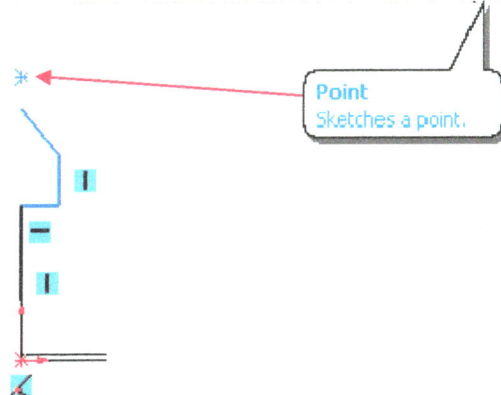

Select the **Add Relation** icon in the CommandManager, or pull down the "Tools" menu and select **Relations – Add**.

Rectangle	Circle	Centerpoint Arc	Tangent Arc	3 Point Arc	Sketch Fillet	Centerline	Spline	Point	Add Relation

Add Relation
Controls the size or position of entities with constraints such as concentric or vertical.

If the point is not already selected, select the point just created. Then, select the first vertical line that you created as shown.

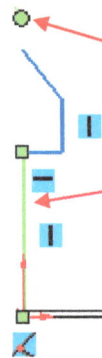

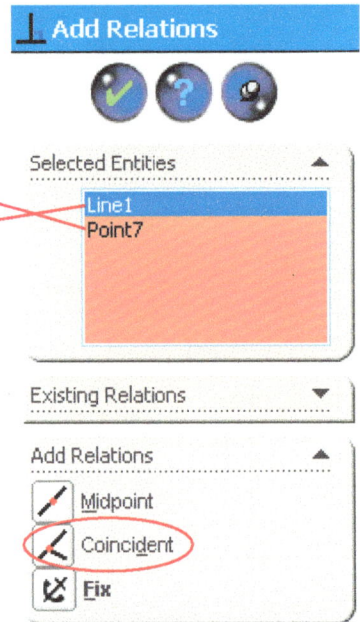

Add Relations

Selected Entities
Line1
Point7

Existing Relations

Add Relations
Midpoint
Coincident
Fix

In the **Add Relations** PropertyManager, select the **Coincident** relation button.

Next, select the endpoint of the angled line and then select the same vertical line again.

Select the **Coincident** relation button again.

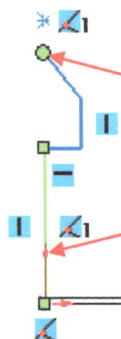

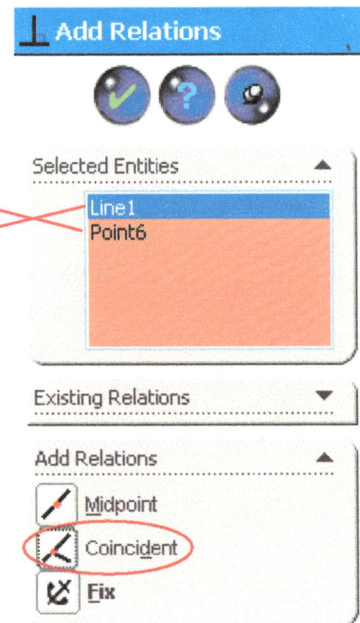

Add Relations

Selected Entities
Line1
Point6

Existing Relations

Add Relations
Midpoint
Coincident
Fix

Then, select the green check mark button at the top of the **Add Relations** PropertyManager.

Select the **Smart Dimension** icon in the CommandManager, or pull down the "Tools" menu and select **Dimensions – Smart**.

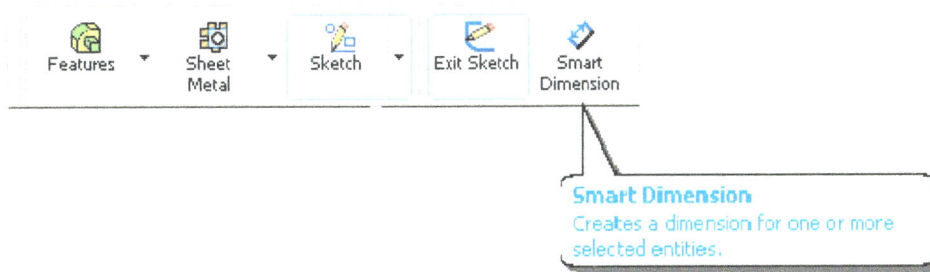

Dimension the sketch as shown to the right by selecting the appropriate line(s) and end points. Make sure you dimension from the origin to the point you created. Also, to add the angle dimension, simply select the angled line, and then select the short vertical line. Moving the cursor around will change the angle measurement. Once you get the correct angle dimension, you can click the right mouse button. This locks in the dimension type and gives you better control over the placement of the dimension.

Don't get upset if the lines move when you add a dimension, changing the shape of your sketch. Remember, all you have to do is press the **Escape** key on the keyboard. Then, drag the endpoint of the line to get the desired shape again. Select the **Smart Dimension** icon again to continue dimensioning.

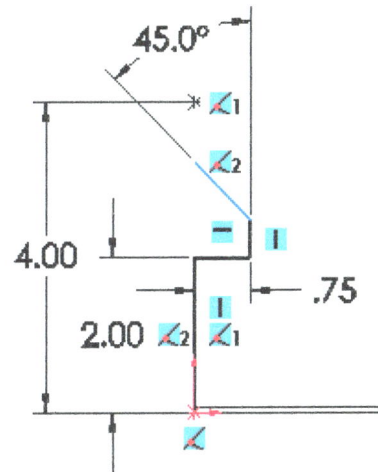

The **4.00** dimension is the control dimension for the overall height. To tie the part to the point, add another dimension from the point to the endpoint of the angled line as shown below.

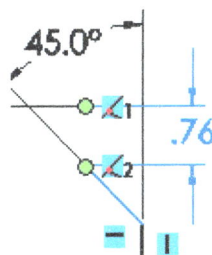

When the **Modify** dialog box appears, pull down the menu and select **Add Equation**.

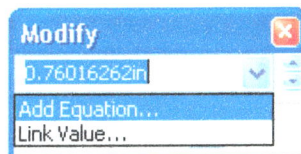

The **Add Equation** dialog box appears and the name of the dimension is already entered for you.

First, type '**Thickness**' and then press the * button. Next, press the **sin** button. Finally select the angle dimension (**45.0°**) in the graphics area. You may need to drag the dialog boxes out of the way in order to see the dimension.

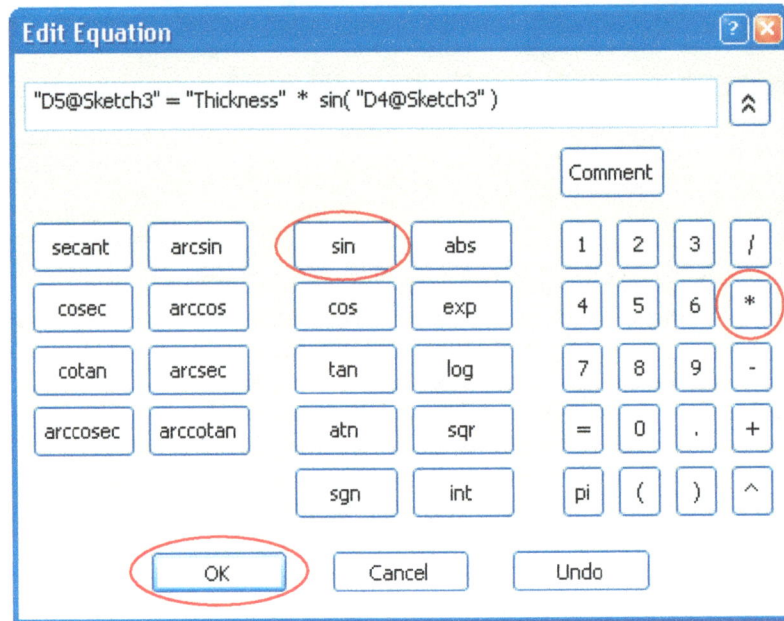

After you have entered the correct equation, select the **OK** button.

Select the **OK** button again to close the **Equations** dialog box.

The dimension is now updated, indicating that it is an equation.

Select the **Exit Sketch** icon in the CommandManager or in the upper right corner of the graphics area.

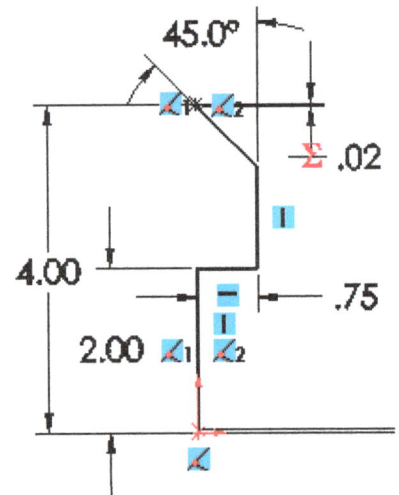

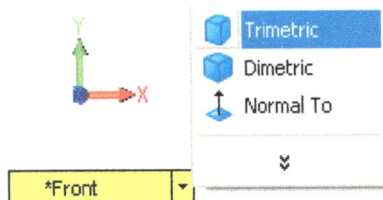

Change back to the **Trimetric** view by selecting the View orientation pull down arrow in the bottom left corner of your graphics area and selecting **Trimetric**.

A preview of the flange will be shown. With the **Miter Flange** feature, you are able to select multiple edge lines to create multiple flanges. Select the remaining three edge lines of the *bottom* face of the base flange. **Note**: If you are having trouble selecting the back bottom line, click the right mouse button and pick **Select Other**.

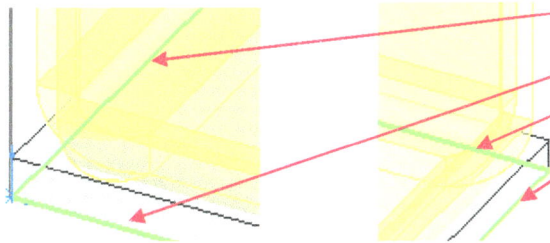

Make sure that the **Flange position** is set to **Material Inside**.

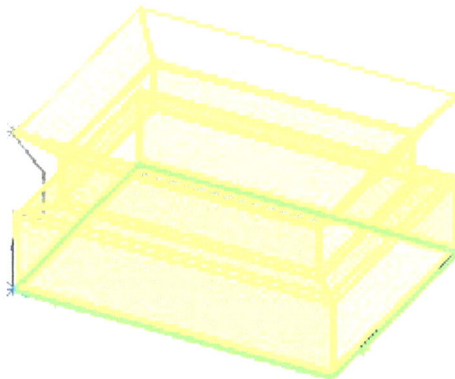

Select the green check mark button at the top of the **Miter Flange** PropertyManager to accept the settings and create the feature.

Showing the Dimensions

In the FeatureManager design tree, right click on **Annotations** and select **Show Feature Dimensions**. You may drag and drop dimensions with the mouse to reposition them and make them easier to see.

If the part is off the screen, select the **Zoom to Fit** icon in the "View" toolbar, or pull down the "View" menu and select **Modify – Zoom to Fit** so you can see the entire part and center it in the graphics area.

Zoom to Fit
Zooms the model to fit the window.

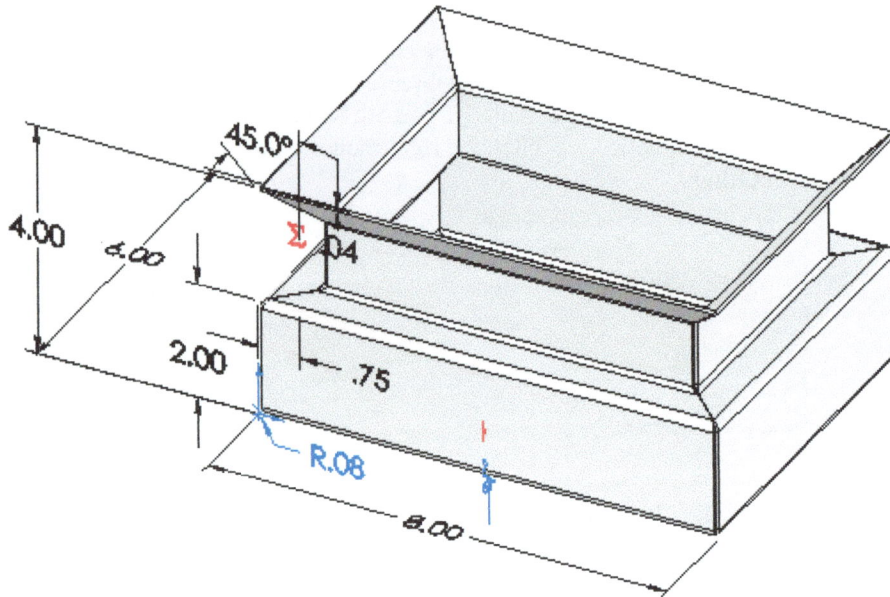

Save the Part

Select the **Save** icon in the "Standard" toolbar, or select **Save** from the "File" pull down menu.

The **Save As** dialog box appears.

In the **File name** box, type the name of the drawing number. For this chapter, use '**68495-188**' and select **Save**.

Changing the Part

Again, if the material thickness, bend radius, length, width, or height dimensions are changed, the part will remain constant, achieving the design intent. The height is also easily updatable because we added the point with an equation. Even if the angle of the top flange changes, the formula ensures that the overall height will remain constant.

Closing the File

Select **Close** from the "File" pull down menu.

Select **No** when prompted to **Save changes to 68495-188.SLDPRT?**

Chapter 4

End Panel

The end panel is a symmetrical part which will use the Mirror command to complete the sketch of the Base Flange. This adds the relationship to the geometry to keep the symmetry, meaning when you add a dimension to one side of the sketch, it is applied to the other side as well.

The Jog command is also used here to create the back and top flanges. To ensure that the top flange holds the same dimension as the side flanges, an equation is added.

When using the Jog command, it is possible that the part will jog in the wrong direction from what you intended. Here you will learn how to make certain the Jog is going your way.

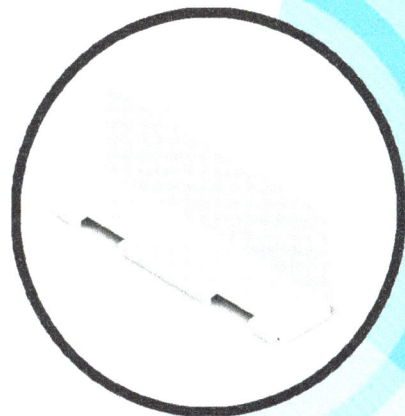

Create a New Part Document

Create a new part document by selecting the **New** icon in the "Standard" toolbar, or pull down the "File" menu and select **New**.

When the **New SolidWorks Document** dialog box appears, select **Part** and then select **OK**.

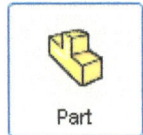

Create the Base Flange

Select the **Sheet Metal** icon in the control area of the CommandManager. Then, select the **Base-Flange/Tab** icon from the toolbar, or pull down the "Insert" menu and select **Sheet Metal – Base Flange**.

The **Front**, **Top**, and **Right** planes appear.

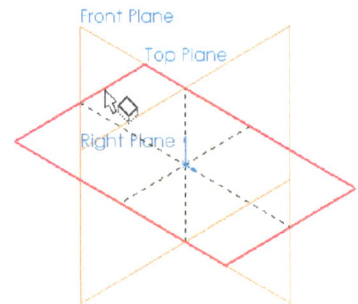

When you are prompted in the PropertyManager to select a plane on which to sketch the feature cross-section, select the **Top Plane** with the cursor.

Select the **Centerline** icon in the CommandManager, or pull down the "Tools" menu and select **Sketch Entities – Centerline**.

Centerline
Sketches a centerline. Use centerlines to create symmetrical sketch elements, revolved features, or as construction geometry.

Create a vertical centerline starting at the origin and ending above the origin.

Select the **Line** icon in the CommandManager, or pull down the "Tools" menu and select **Sketch Entities – Line**.

Create the profile shown to the right of the centerline you just created. You will dimension the sketch in the next section. The shape of the sketch is all that is important right now. If you create a line that you do not want, press the **Escape** key. Then, pick the line you do not want and press the **Delete** key.

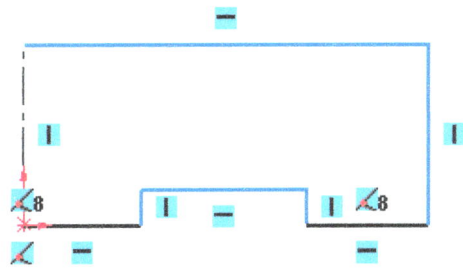

Select the **Line** icon in the CommandManager again to deselect the **Line** tool, or pull down the "Tools" menu and select **Sketch Entities – Line**. You can also press the **Escape** key.

Select all the geometry by pressing the left mouse button above and to the left of the sketch geometry and dragging the green selection box down and to the right around the geometry in the graphics area.

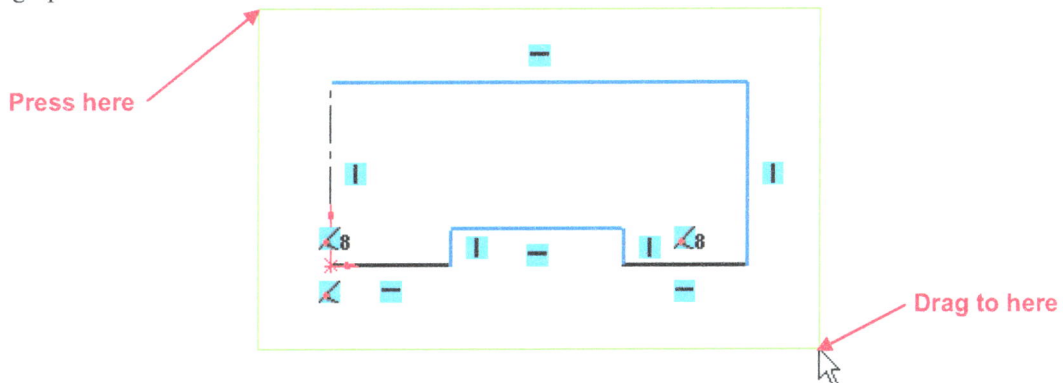

Press here

Drag to here

With all the geometry highlighted, select the **Mirror** icon in the CommandManager, or pull down the "Tools" menu and select **Sketch Tools – Mirror**. The **Mirror** icon in the CommandManager may be off the screen. In order to show the additional icons, simply select the double chevrons (>>) at the right edge of the CommandManager.

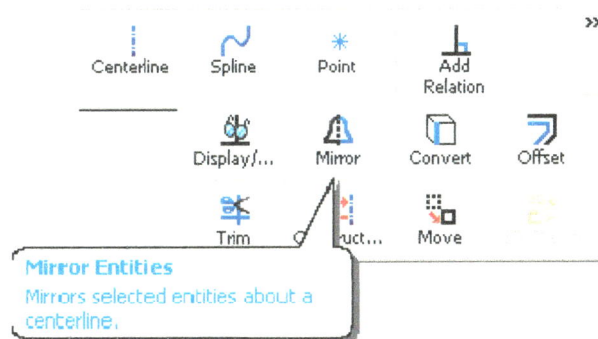

Mirror Entities
Mirrors selected entities about a centerline.

The sketch is mirrored to the other side of the centerline.

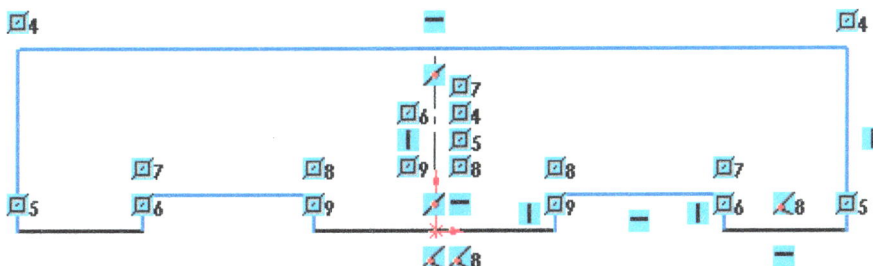

Adding Dimensions

Select the **Smart Dimension** icon in the CommandManager, or pull down the "Tools" menu and select **Dimensions – Smart**.

Select the vertical right side line of the shape and place a '**4.00**' dimension to the right of the line.

Select the top horizontal line of the shape and place a '**15.00**' dimension above the line.

Select the **Zoom to Fit** icon in the "View" toolbar, or pull down the "View" menu and select **Modify – Zoom to Fit** so you can see the entire rectangle and center it in the graphics area. You may also press the **F** key on the keyboard, the keyboard shortcut for **Zoom to Fit**.

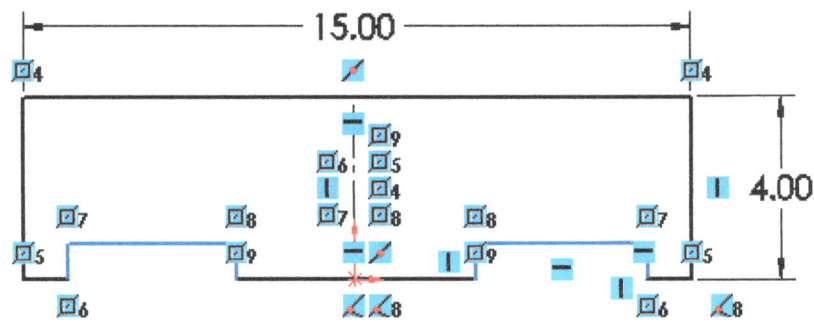

Select the bottom right horizontal line of the shape and place a '**2.00**' dimension. Then, place a '**6.00**' dimension below the bottom middle line. Finally, place a '**0.50**' dimension of the small vertical line at the bottom right of the shape.

If you are having trouble picking the line (red dots appear and you keep picking the endpoint), hold down the **Shift** key on the keyboard and then press the **Z** key on the keyboard to use the shortcut for the **Zoom in** command to enlarge the notch area. (**Zoom in** keyboard shortcut: Shift+z. Zoom out: z)

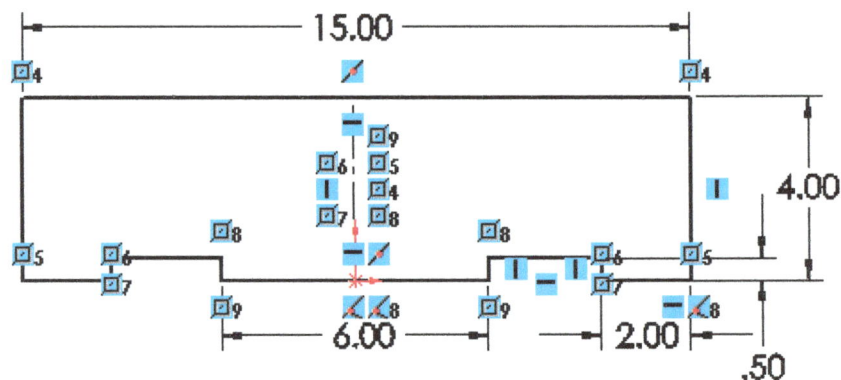

Select the **Exit Sketch** icon in the CommandManager or in the upper right corner of the graphics area.

Complete the Base Flange

In the **Base Flange** PropertyManager under **Sheet Metal Parameters**, set the **Thickness** to '**0.075in**'.

Check the **Reverse direction** check box to make sure that the preview of the extrusion is in the Y-positive direction (The yellow rectangle will be above the green rectangle).

Select the green check mark button at the top of the **Base Flange** PropertyManager to accept the settings and create the feature.

Create the Front Flanges

Select the **Edge Flange** icon in the CommandManager, or pull down the "Insert" menu and select **Sheet Metal – Edge Flange**.

Select the left most front edge. A preview of the flange will be shown. Move the cursor down and click to set the direction of this flange.

Select the other two front edges as shown to create a total of three edge flanges in one feature.

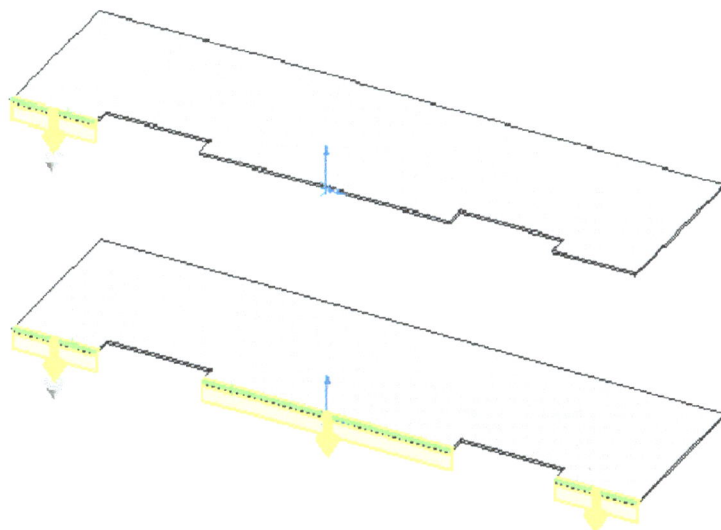

In the **Edge-Flange** PropertyManager, set the **Flange Length** to a **Blind Length** of '**0.50in**'.

Make sure that the **Outer Virtual Sharp** button is depressed.

Set the **Flange Position** to **Material Inside**.

Select the green check mark button at the top of the **Edge-Flange** PropertyManager to create the front flanges.

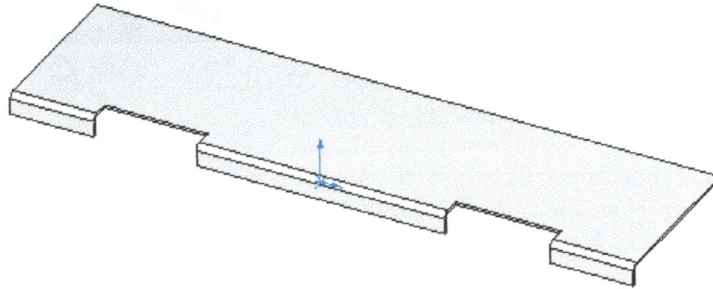

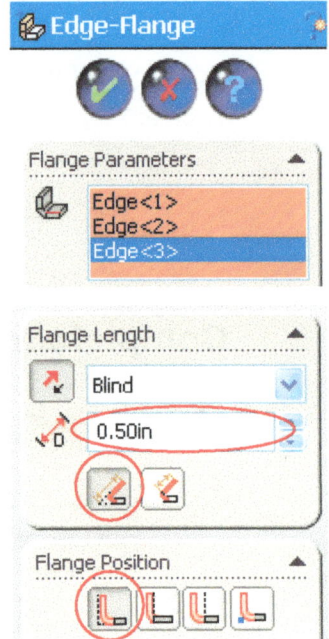

Add a Jog

Select the **Jog** icon in the CommandManager, or pull down the "Insert" menu and select **Sheet Metal – Jog**.

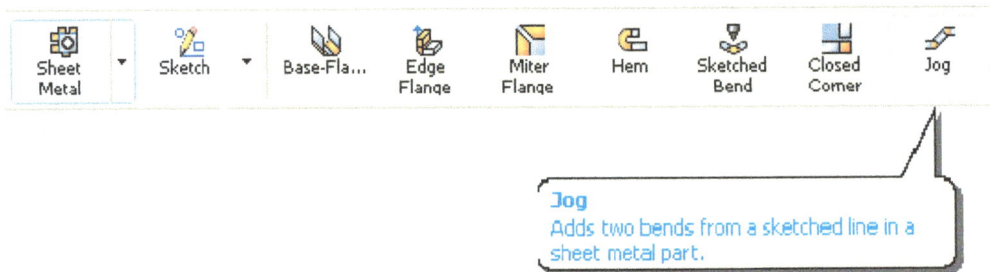

Jog
Adds two bends from a sketched line in a sheet metal part.

When prompted in the PropertyManager to select a planer face on which to sketch bend lines, pick the top of the part. When you select the planer face, a sketch opens on that plane.

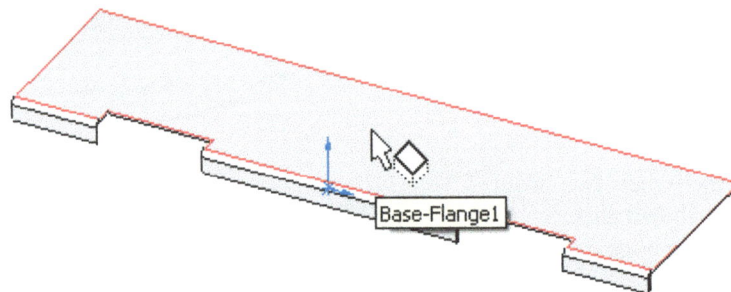

Select the **Line** icon in the CommandManager, or pull down the "Tools" menu and select **Sketch Entities – Line**.

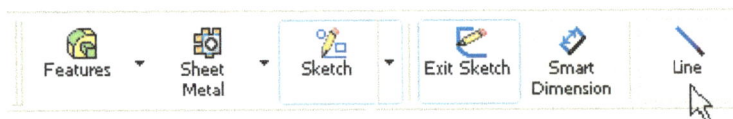

SolidWorks for the Sheet Metal Guy

In the **Insert Line** PropertyManager, set the **Orientation** to **Horizontal** to add a horizontal relation to the line. The **Jog** feature does not require a horizontal or vertical line but it must contain only one line. The line doesn't need to be the full length of the flange, but it's best not to exceed the edges of the part.

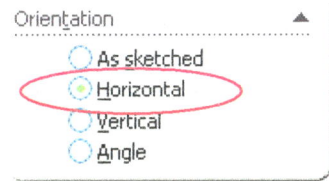

Orientation
- As sketched
- Horizontal
- Vertical
- Angle

Create a horizontal line on the top of the part as shown.

To show that the line is horizontal, the cursor changes to

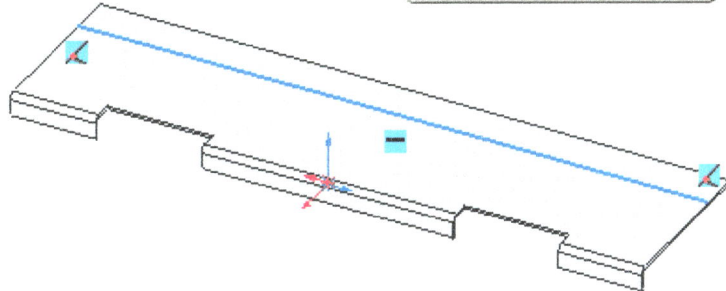

Select the **Smart Dimension** icon in the CommandManager, or pull down the "Tools" menu and select **Dimensions – Smart**.

Features   Sheet Metal   Sketch   Exit Sketch   Smart Dimension

Select the front bend line and the created line to dimension from the front of the part to the line.

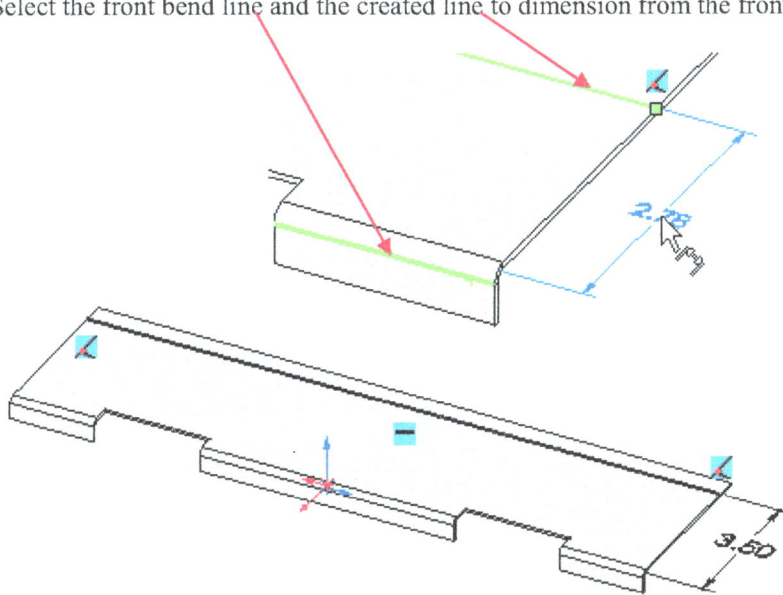

Place the dimension and change the value to '**3.50**'.

Select the **Exit Sketch** icon in the CommandManager or in the upper right corner of the graphics area.

Features   Sketch   Exit Sketch   Smart Dimension

Select the top face again for the fixed face. A large black dot will appear where you select on the part. Make sure the selection point is below the horizontal line.

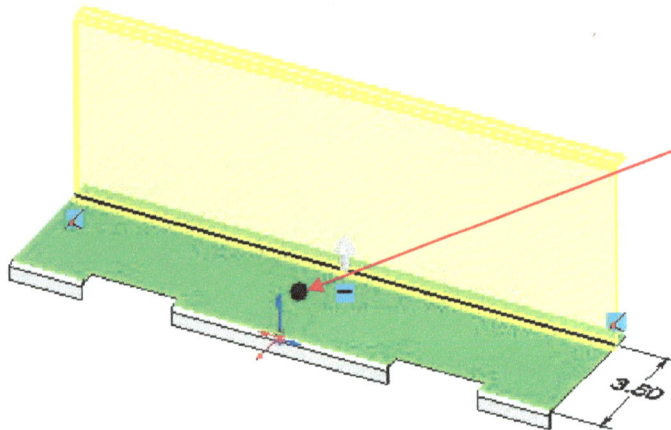

Jog

Selections

Fixed Face:

Face<1>

☑ Use default radius

0.03in

Jog Offset

Blind

D1 5.00in

Dimension position:

☑ Fix projected length

Jog Position

Jog Angle

90.00deg

D1   Set the **Jog Offset** to a **Blind** distance of **5.00in**.

Make sure the **Outside Offset Dimension position** button is depressed and that the **Fix projected length** check box is checked.

Also, make sure that the **Material Outside Jog Position** button is depressed and that the **Jog Angle** is set to **90.00deg**.

Select the green check mark button at the top of the **Jog** PropertyManager to accept the settings and create the feature.

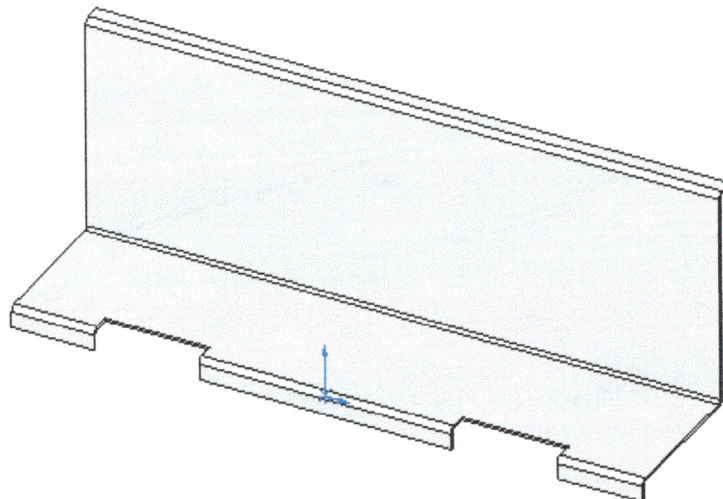

SolidWorks for the Sheet Metal Guy

Create the Side Flanges

Select the **Edge Flange** icon in the CommandManager, or pull down the "Insert" menu and select **Sheet Metal – Edge Flange**.

Select the right most bottom edge as shown. Pull down the preview flange and click the left mouse button to set the flange direction.

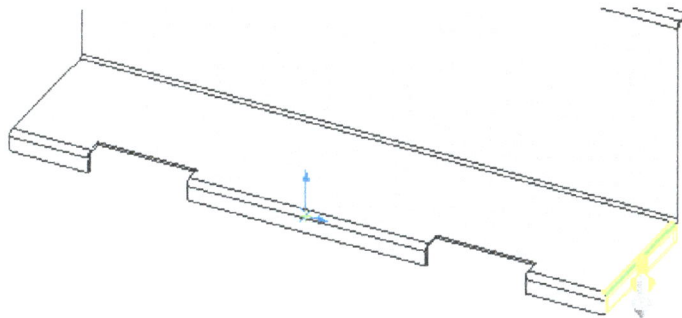

Select the other three edges of the part as shown.

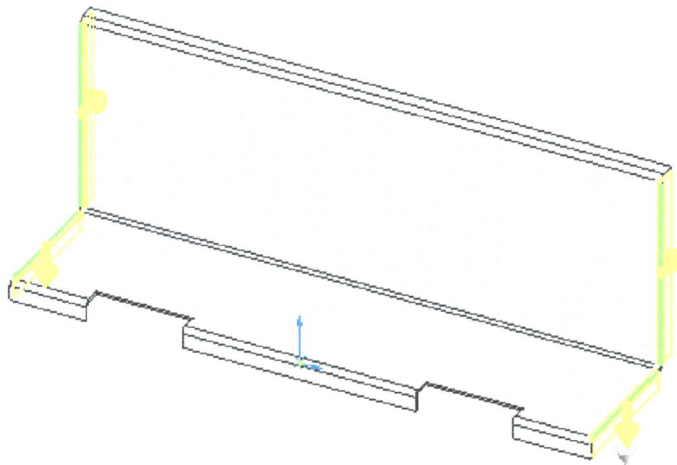

Set the following settings: **Flange Length** to a **Blind** distance of '**0.50in**', depress the **Outer Virtual Sharp** button and the **Material Inside Flange Position** button, and check the **Trim side bends** check box.

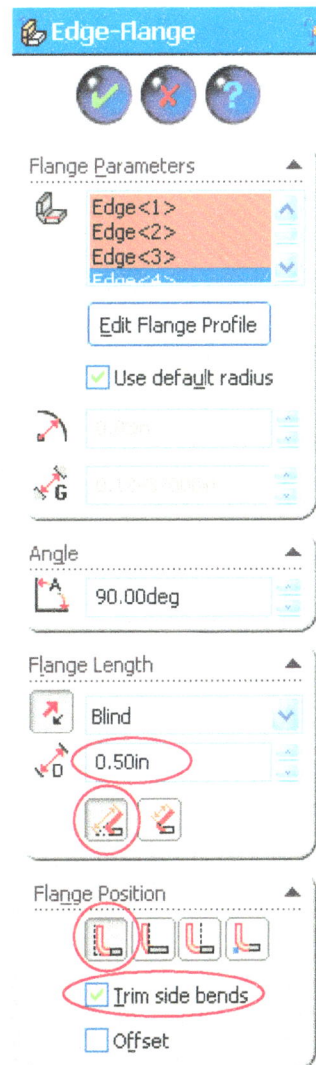

Select the green check mark button at the top of the **Edge-Flange** PropertyManager to accept the settings and create the side flanges.

Add Corner Breaks

Select the **Break-Corner/Corner-Trim** icon in the CommandManager, or pull down the "Insert" menu and select **Sheet Metal – Break Corner**.

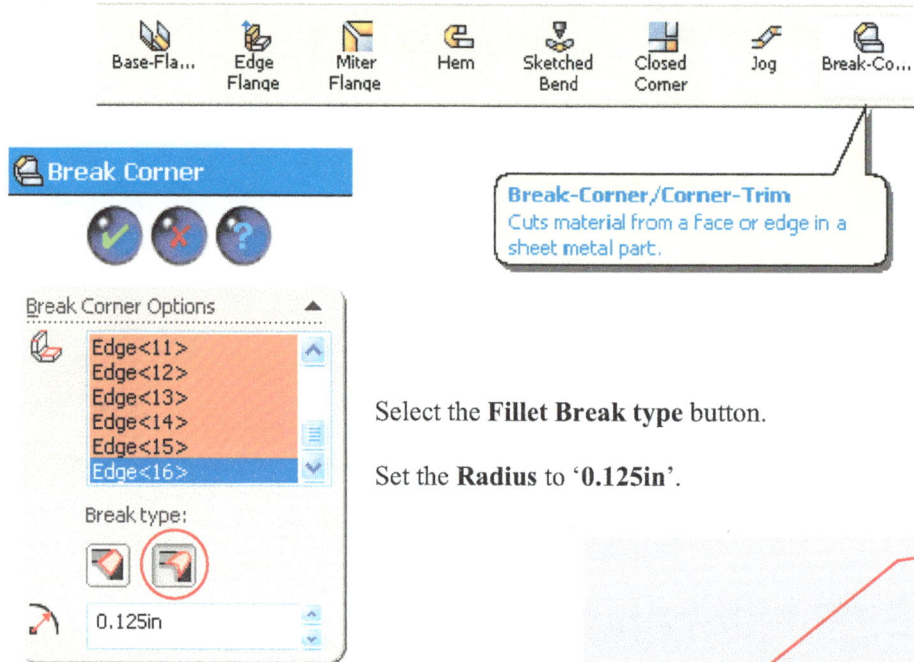

Select the **Fillet Break type** button.

Set the **Radius** to '**0.125in**'.

Next, select the corner edge line of each of the flange corners as shown.

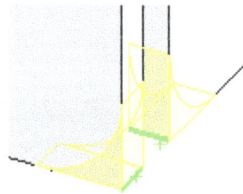

Select these corners.

Do the same on the left side of the part as well.

To rotate the sheet metal model around so that you can easily select the appropriate edges, you may use the **Rotate View** icon from the View toolbar, or pull down the "View" menu and select **Modify – Rotate**. Remember that the middle mouse button is a shortcut to access the **Rotate View** tool. Simply drag using the middle mouse button to rotate the view. When you release the middle mouse button, the tool is automatically deactivated.

Press the left mouse button and drag the cursor to rotate the view.

Select the **Rotate View** icon from the View toolbar again, or pull down the "View" menu and select **Modify – Rotate** in order to deactivate the tool.

Select the green check mark button at the top of the **Break Corner** PropertyManager to accept the settings and create the feature.

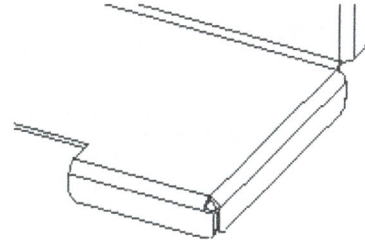

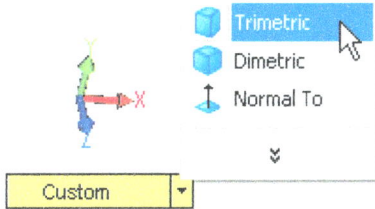

Change the display view to **Trimetric** by selecting **Trimetric** in the View list in the lower left corner of the graphics area.

Show the Dimensions

In the FeatureManager design tree, right click on **Annotations** and select **Show Feature Dimensions**.

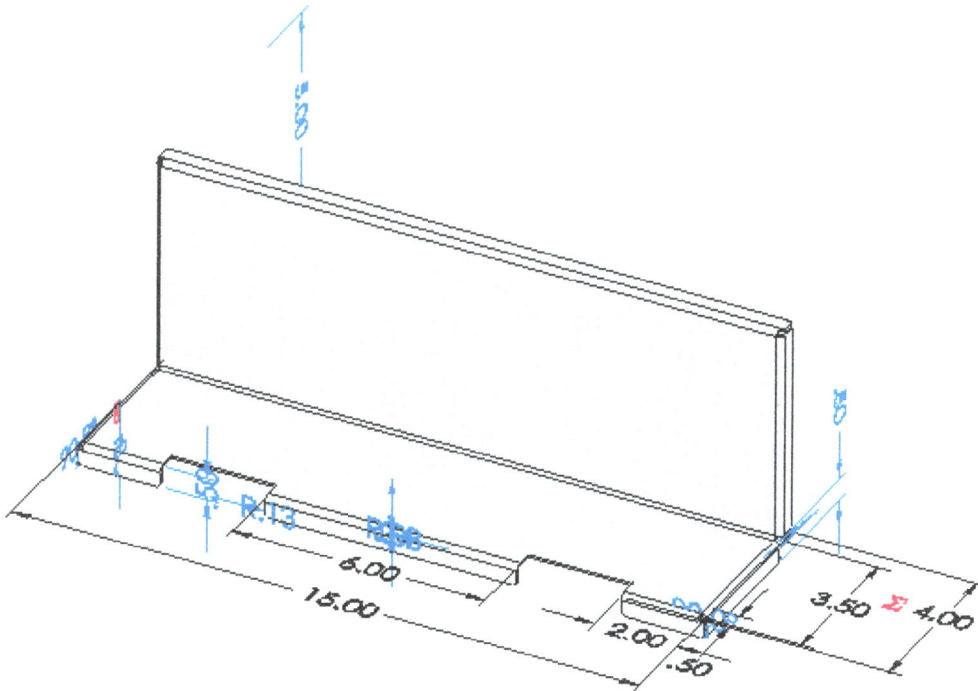

Saving the Part

Select the **Save** icon in the "Standard" toolbar, or select **Save** from the "File" pull down menu.

The **Save As** dialog box appears.

In the **File name** box, type the name of the drawing number. For this lesson, use '**20636-26**' and select **Save**.

Fixing a Jog Feature Error

Start by breaking the part, or shall we say creating the error. Double click on the **3.50** dimension and change the value to '**1.0**.'

Select the **Rebuild** icon in the "Standard" toolbar or press the keyboard shortcut **Ctrl+B**.

If the **What's Wrong** dialog box appears, select the **Close** button. If the **What's Wrong** dialog box appeared, the part is now facing in the wrong direction. If your part looks correct just continue reading to explain why.

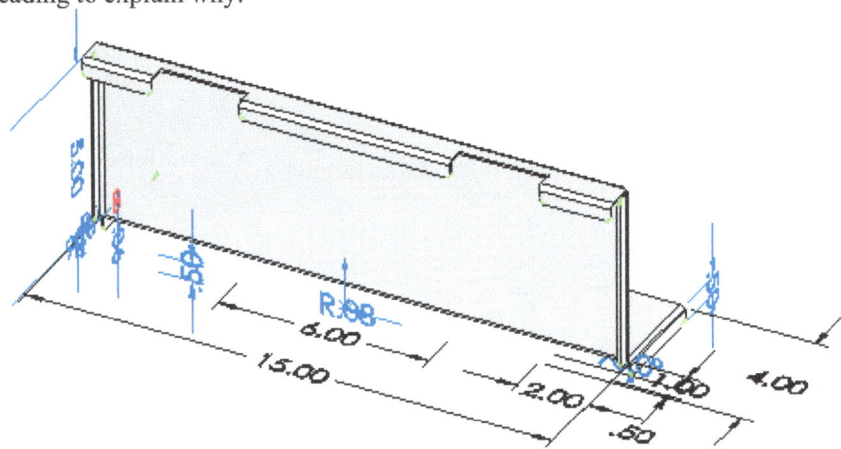

To fix this, right-click on **Jog1** in the FeatureManager design tree and select **Edit Feature**.

Look closely at the jog preview and pay attention particularly to the large black dot that indicates your selection point when you created the jog. The selection point is now above the Jog sketch line instead of below it.

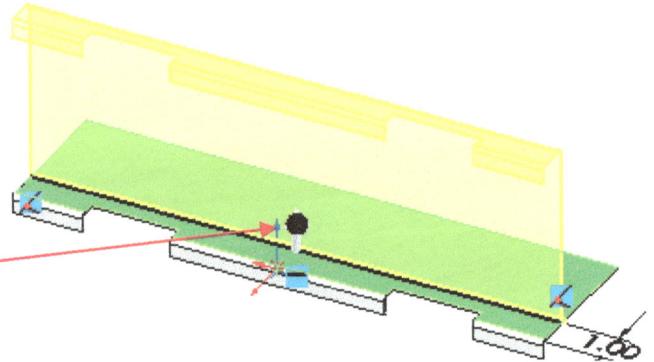

To fix this problem, simply click once on the top face to deselect it, and then click again below the horizontal line to reselect the top face and reposition the selection point. The preview should now look correct.

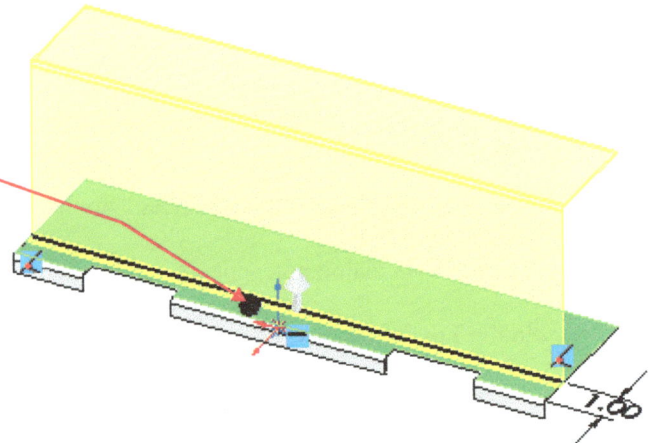

Select the green check mark button at the top of the **Jog1** PropertyManager to accept the new settings and recreate the part.

Holding the Proper Dimensions

With the Jog dimension modified, the top flange is no longer **0.50in**. The **4.00** dimension is what is controlling the width of the part, and in turn the width of the flange. An equation can be used to make sure that the top flange width will be the same width as the side flanges, even if the part is modified.

To do this, double click on the **4.00** dimension. If you do not see all the dimensions, select the **Rebuild** icon in the "Standard" toolbar or press the keyboard shortcut **Ctrl+B**.

When the **Modify** dialog box appears, pull down the menu and select **Add Equation**.

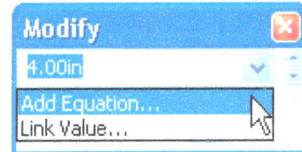

In the graphics, select the **1.00** dimension. Then, select the plus (+) button in the **Add Equation** dialog box. Finally, select the **.50** (flange length) vertical dimension.

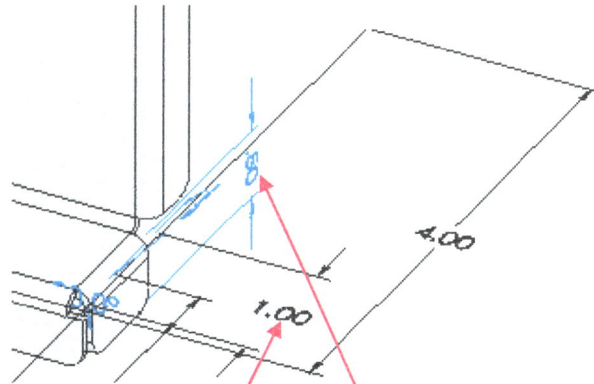

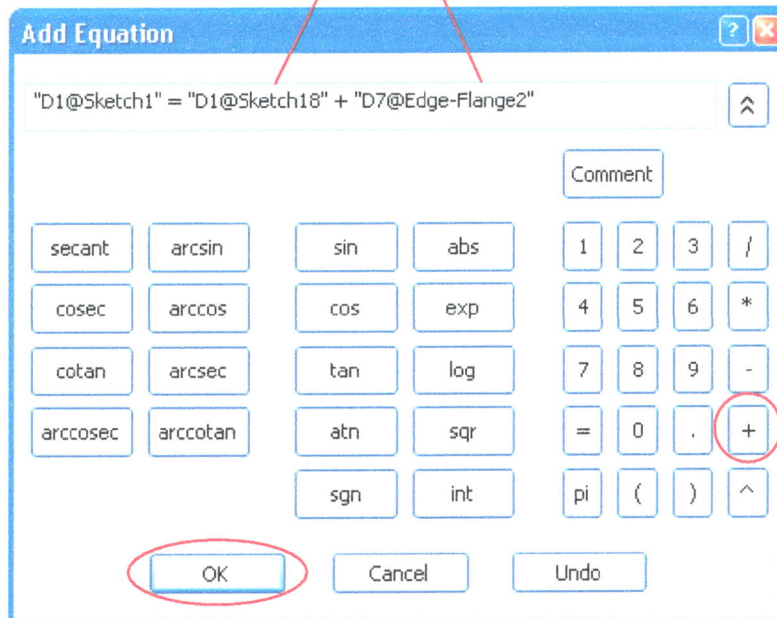

Select the **OK** button in the **Add Equation** dialog box and in the **Equations** dialog box.

The dimension will change to

Select the **Rebuild** icon in the "Standard" toolbar or press the keyboard shortcut **Ctrl+B**.

The part is now correct, holding your desired dimensions. To verify this, double click on the **1.00** dimension and change it back to **3.5** and select the **Rebuild** icon. Now the overall width dimension automatically updates, holding the top flange width equal to the side flange width.

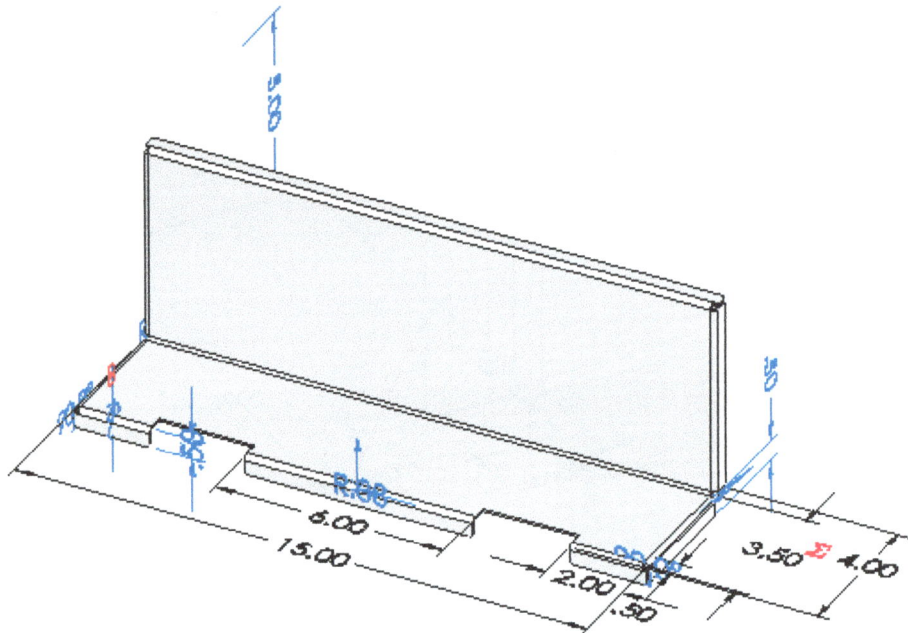

Changing the Part

Again, if the material thickness, bend radius, length, width, height dimensions are changed, the key dimensions will remain constant, achieving the design intent.

Closing the File

Select **Close** from the "File" pull down menu.

Select **Yes** when prompted to **Save changes to 20636-26.SLDPRT?** because you have modified the part since it was last saved.

Chapter 5

Bracket

This bracket provides the opportunity to use several of the flange commands, showing their versatility. The split in the two back flanges poses a situation where you must ensure the flat pattern is manufacturable.

The Base Flange and Edge Flanges of this part are not simple rectangles. Therefore, additional editing is required to create the flange shapes.

The Jog command is again used to create an offset on one of the back flanges.

The Hem command is introduced. SolidWorks provides a very flexible Hem command which makes it easy to adjust to the desired results.

Create a New Part Document

To begin this chapter, create a new part document by selecting the **New** icon in the "Standard" toolbar, or pull down the "File" menu and select **New**.

When the **New SolidWorks Document** dialog box appears, select **Part** and then select **OK**.

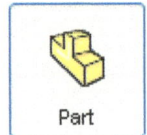

Create the Base Flange

Select the **Sheet Metal** icon in the control area of the CommandManager. Then, select the **Base-Flange/Tab** icon from the toolbar, or pull down the "Insert" menu and select **Sheet Metal – Base Flange**.

When you are prompted in the PropertyManager to select a plane on which to sketch the feature cross-section, select the **Top Plane** with the cursor.

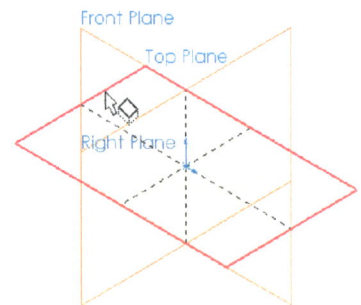

Create a Rectangle

Select the **Rectangle** icon in the CommandManager, or pull down the "Tools" menu and select **Sketch Entities – Rectangle**.

Click on the origin to start the rectangle. Move the cursor up and to the right. Click again to complete the rectangle.

Select the **Smart Dimension** icon in the CommandManager, or pull down the "Tools" menu and select **Dimensions – Smart**.

Select the left vertical line of the rectangle and place a '**5.00**' dimension. Then, select the bottom horizontal line and place a '**12.00**' dimension.

Press the **F** key on the keyboard, the keyboard shortcut for **Zoom to Fit**, so you can see the entire rectangle and center it in the graphics area.

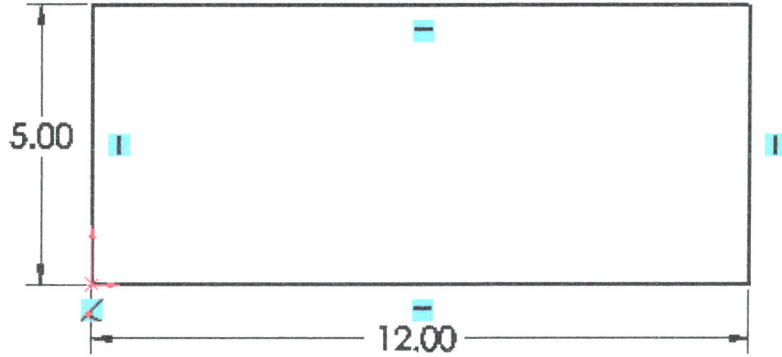

Select the **Line** icon in the CommandManager, or pull down the "Tools" menu and select **Sketch Entities – Line**.

Create a **1.00** wide by **4.00** long cutout in the upper right corner of the rectangle, as shown below, by creating two lines and dimensioning them accordingly.

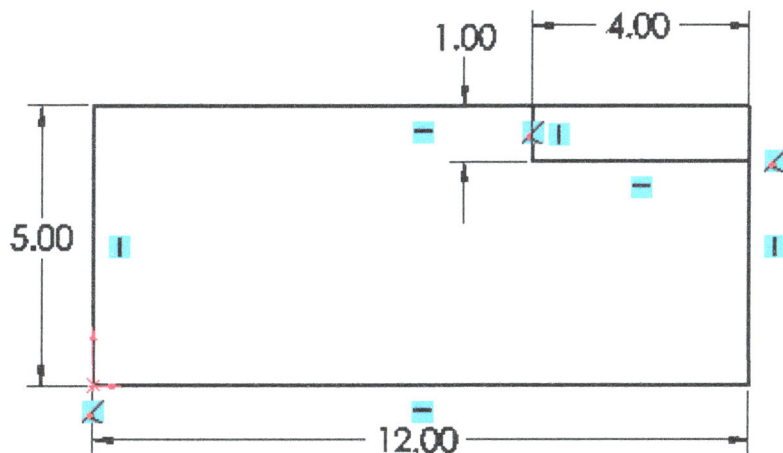

Select the **Trim** icon in the CommandManager, or pull down the "Tools" menu and select **Sketch Tools – Trim**. The **Trim** icon may be off the screen. In order to show the additional icons, simply select the double chevrons (>>) at the right edge of the CommandManager.

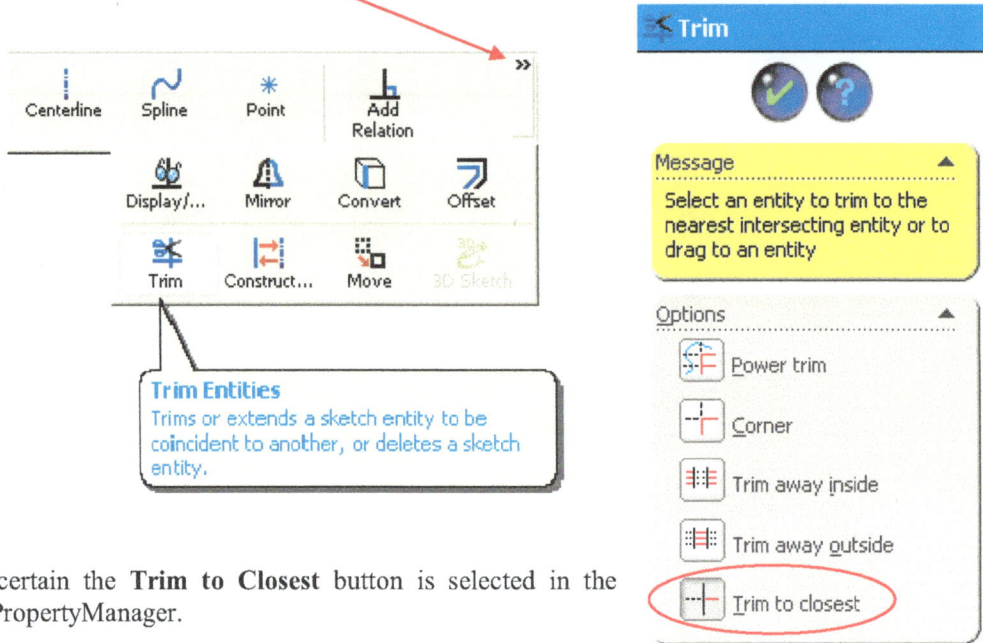

Make certain the **Trim to Closest** button is selected in the **Trim** PropertyManager.

Select the portion of the top horizontal line that you want to trim (throw away) and click the left mouse button.

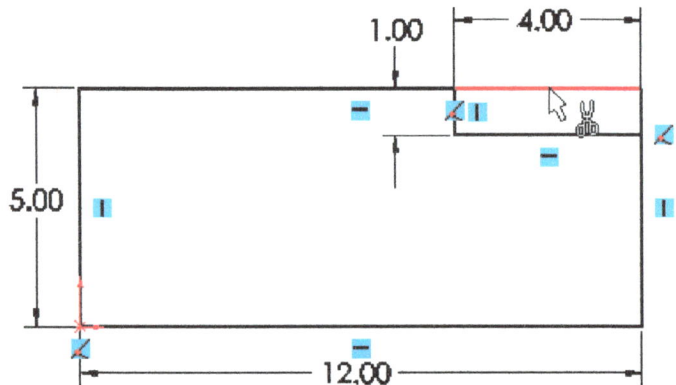

Select the portion of the right vertical line that you want to trim (throw away) and click the left mouse button.

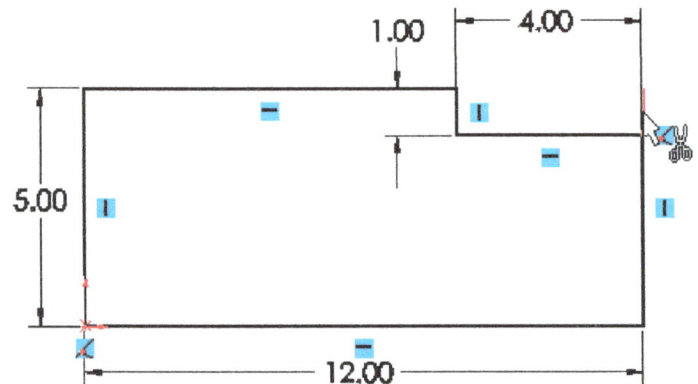

You can use the **Undo** command if you don't get the desired results by selecting the **Undo** icon in the "Standard" toolbar, or pull down the "Edit" menu and select **Undo Trim Entities**.

Undo
Reverses the last action.

Select the **Exit Sketch** icon in the CommandManager or in the upper right corner of the graphics area.

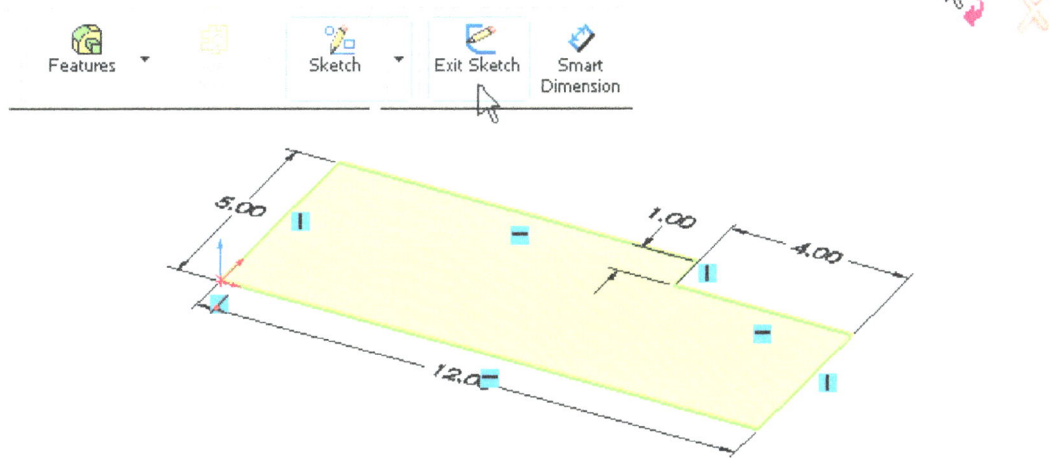

Features   Sketch   Exit Sketch   Smart Dimension

5.00   1.00   4.00
12.0

Completing the Base Flange

Base Flange

In the **Base Flange** PropertyManager under **Sheet Metal Parameters**, set the **Thickness** to '**0.059in**'.

Check the **Reverse direction** check box to make sure that the preview of the extrusion is in the Y-positive direction (The yellow rectangle will be above the green rectangle).

Select the green check mark button at the top of the **Base Flange** PropertyManager to accept the settings and create the base flange.

Sheet Metal Gauges
☐ Use gauge table

Sheet Metal Parameters
T1   0.059in

☑ Reverse direction

Create the Side Flange

Select the **Edge Flange** icon in the CommandManager, or pull down the "Insert" menu and select **Sheet Metal – Edge Flange**.

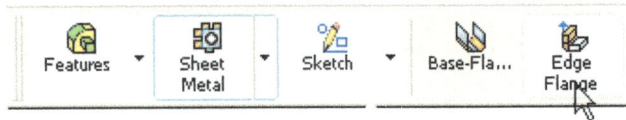

Select the back left edge as the edge to add the flange to.

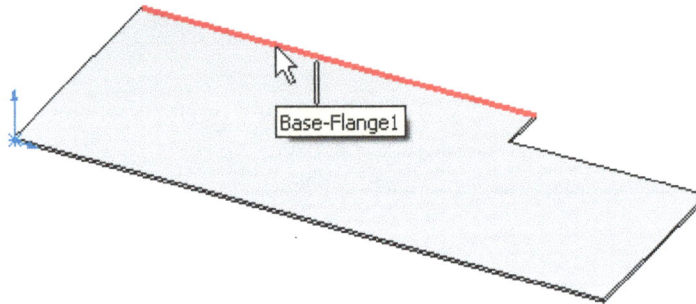

A preview of the flange will be shown. Move the cursor up and click to set the direction of the flange upward.

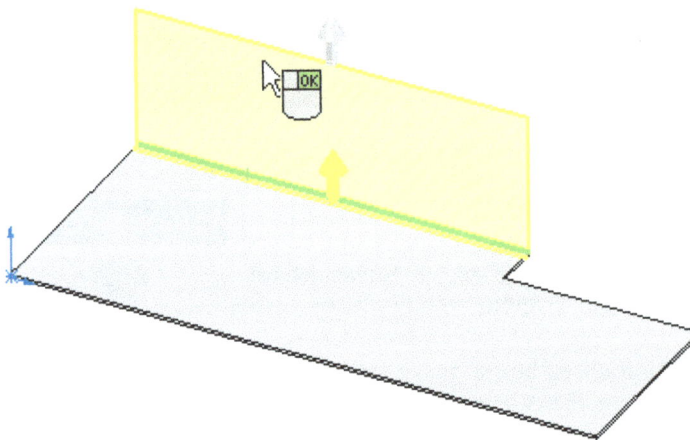

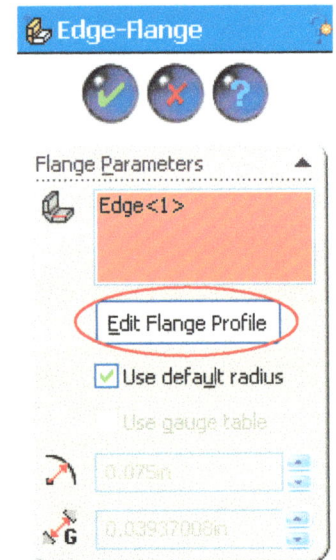

In the **Edge-Flange** PropertyManager, select the **Edit Flange Profile** button. The **Profile Sketch** dialog box will appear. Don't pick anything in this dialog box. Just ignore it for now.

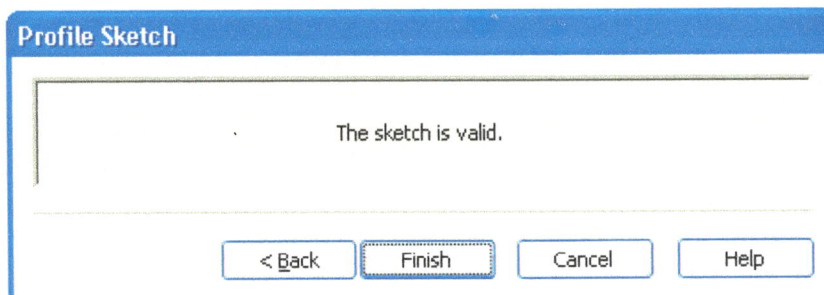

In the lower left corner of the graphics area, select the View list and select **Normal To**. This changes the display view to be **Normal To** the sketch.

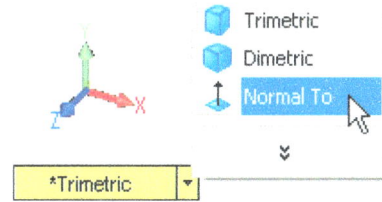

Select the **Line** icon in the CommandManager, or pull down the "Tools" menu and select **Sketch Entities – Line**.

Create the following three lines in the sketch. After creating the first two lines on the left of the sketch, select the **Line** icon twice to stop and restart the line command. Then, create the angled line on the right side of the sketch.

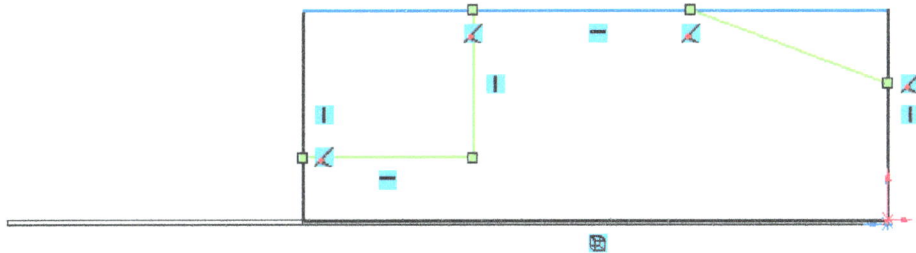

Select the **Trim** icon in the CommandManager, or pull down the "Tools" menu and select **Sketch Tools – Trim**.

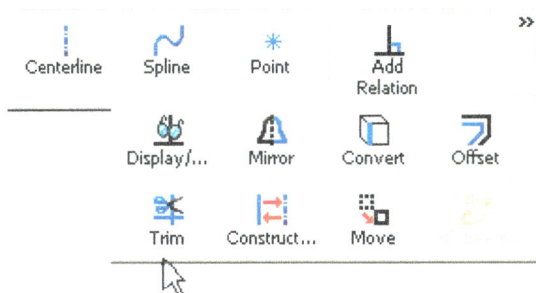

Trim the lines as shown below by selecting the portion of the line that you want to trim (delete) and click the left mouse button. If you see the warning: "The sketch segment being trimmed has a midpoint or equal length relation. Trimming the segment will delete the relation. Do you want to continue?", select the **Yes** button.

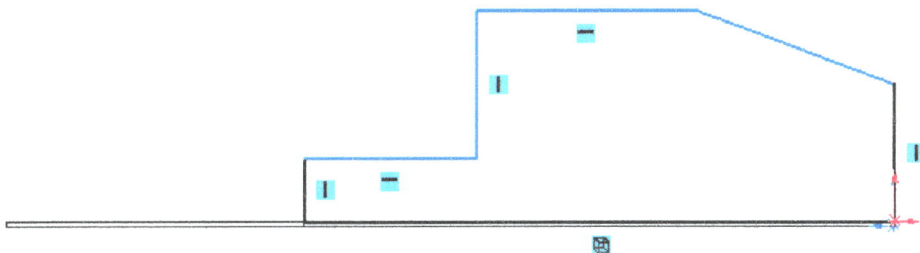

Select the **Smart Dimension** icon in the CommandManager, or pull down the "Tools" menu and select **Dimensions – Smart**.

Dimension the sketch as shown below. Take note that the vertical dimensions are dimensioned from the outside of the thickness. To create the angle dimension, select the right vertical line followed by the angled line. Move the cursor to get the inner angle as shown.

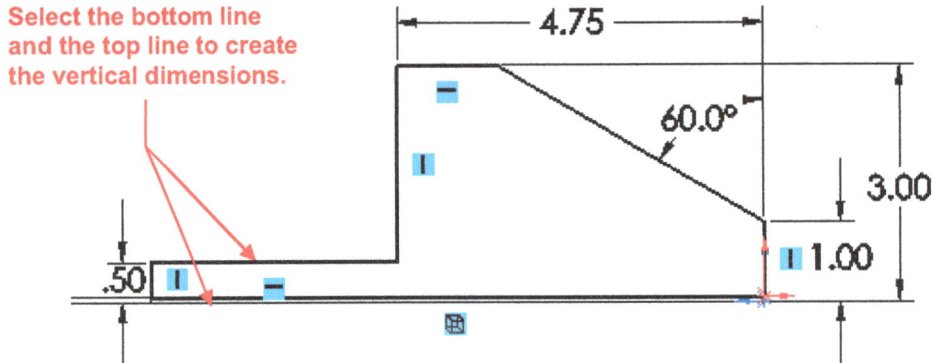

Select the bottom line and the top line to create the vertical dimensions.

4.75

60.0°

3.00

1.00

.50

When you have finished dimensioning the sketch, select the **Back** button in the **Profile Sketch** dialog box.

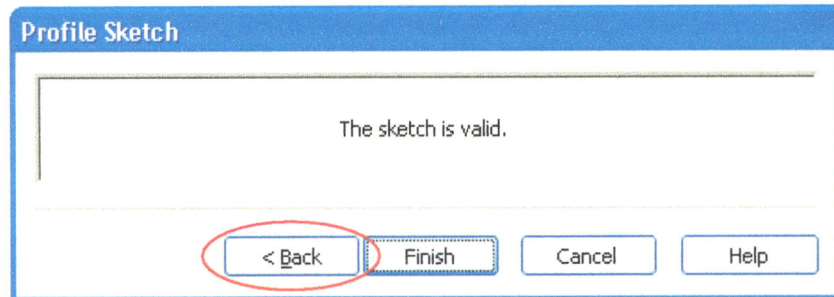

Profile Sketch

The sketch is valid.

< Back | Finish | Cancel | Help

In the **Edge-Flange** PropertyManager, select the **Material Inside Flange Position** button.

Select the green check mark button at the top of the **Edge-Flange** PropertyManager to accept the settings and create the feature.

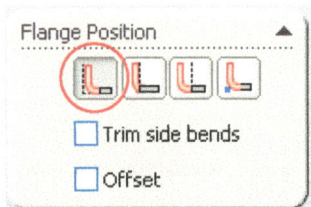

Flange Position

☐ Trim side bends

☐ Offset

Select the **Previous View** icon in the "View" toolbar, or pull down the "View" menu and select **Modify – Previous View** to return to the Trimetric view.

Previous View
Displays the previous view.

Creating a Second Flange

Select the **Edge Flange** icon in the CommandManager, or pull down the "Insert" menu and select **Sheet Metal – Edge Flange**.

Select the back right edge as the edge where you want to add the new flange.

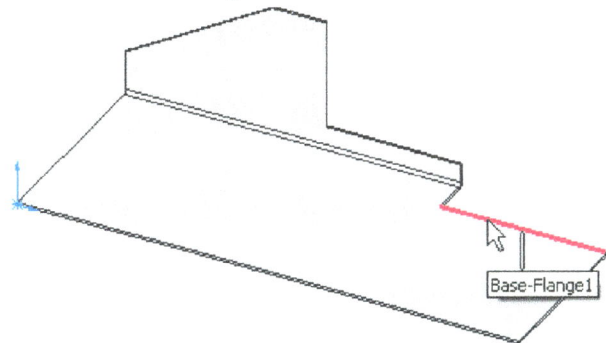

A preview of the flange will be shown. Move the cursor up and click to set the direction of the flange.

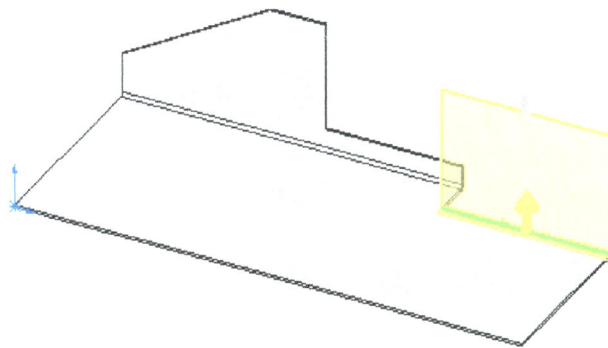

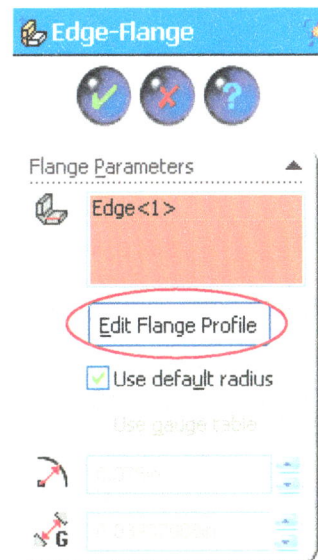

In the **Edge-Flange** PropertyManager, select the **Edit Flange Profile** button.

Again the **Profile Sketch** dialog will appear. Do not pick anything in this dialog box. Again, you will ignore it for right now.

Change the display view to **Normal To** the sketch by selecting **Normal To** in the "View" pop-up menu in the lower left corner of the graphics area.

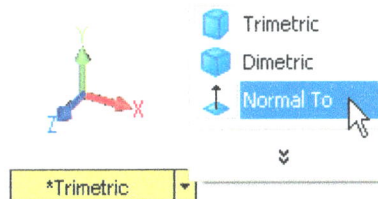

Left click on the right vertical line and drag it to the left. The actual distance does not matter, as this will be controlled by a dimension.

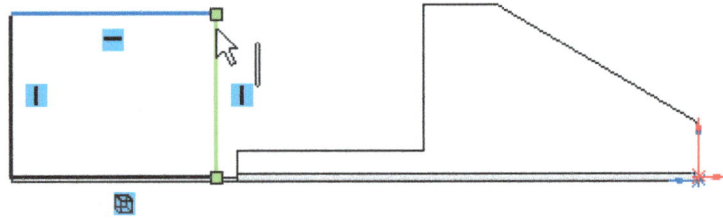

Select the **Line** icon in the CommandManager, or pull down the "Tools" menu and select **Sketch Entities – Line**.

Create the two lines as shown below.

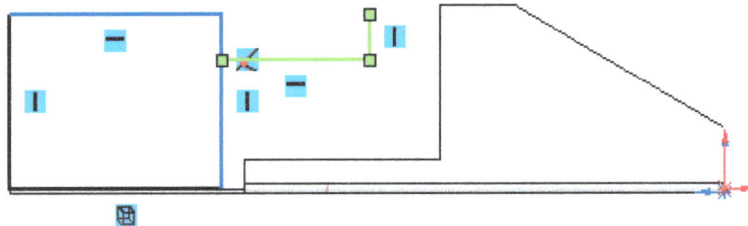

Select the **Trim** icon in the CommandManager, or pull down the "Tools" menu and select **Sketch Tools – Trim**.

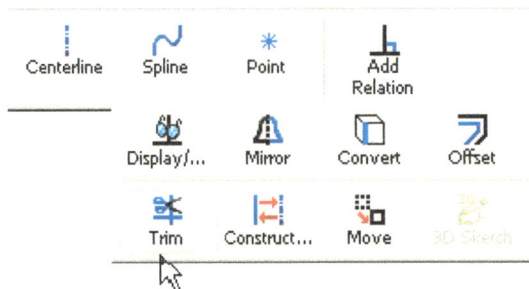

Trim the top portion of the vertical line as shown.

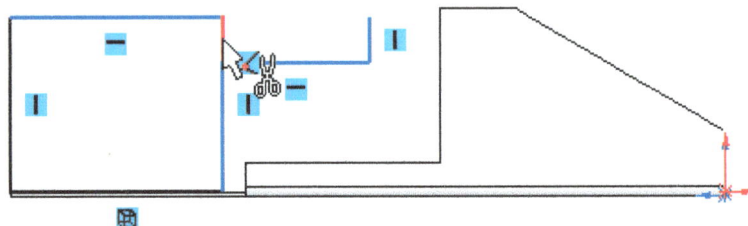

Select the green check mark button at the top of the **Trim** PropertyManager to close the dialog.

Move the cursor to the right end of the top horizontal line. A red dot will appear when you are there and the cursor will change as shown.

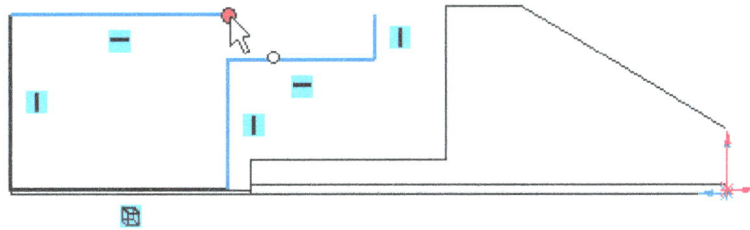

Left click and drag the endpoint to the endpoint of the small vertical line on the right to connect the sketch lines. The cursor will change as shown.

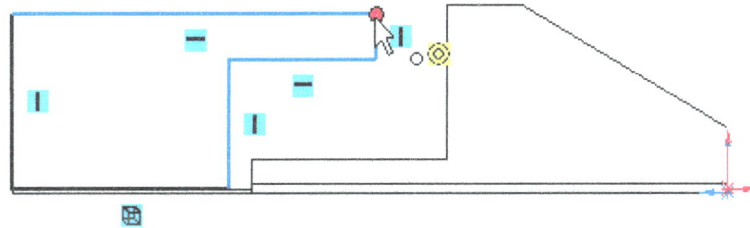

To complete the sketch, select the **Smart Dimension** icon in the CommandManager, or pull down the "Tools" menu and select **Dimensions – Smart**.

Dimension the sketch as shown below. For the vertical dimensions, dimensions from the bottom edge line to maintain the outside dimension. If you are having trouble selecting the bottom line, remember that you can right click near the line and **Select Other**. The precision of an individual dimension can be changed by selecting the dimension and changing the **Tolerance/Precision** in the **Dimension** PropertyManager. The precision of the **.125** dimension was changed from two decimal places to three decimal places (**.123**).

Select the **Back** button in the **Profile Sketch** dialog box.

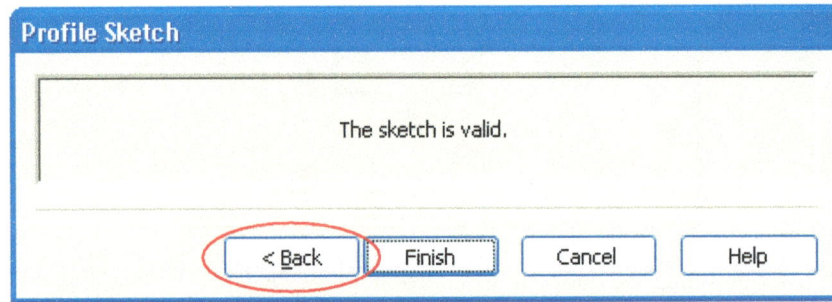

In the **Edge-Flange** PropertyManager, select the **Bend Outside Flange Position** button.

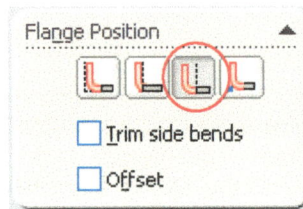

Select the green check mark button at the top of the **Edge-Flange** PropertyManager to accept the settings and create the feature.

Select the **Previous View** icon in the "View" toolbar, or pull down the "View" menu and select **Modify – Previous View** to return to the Trimetric view.

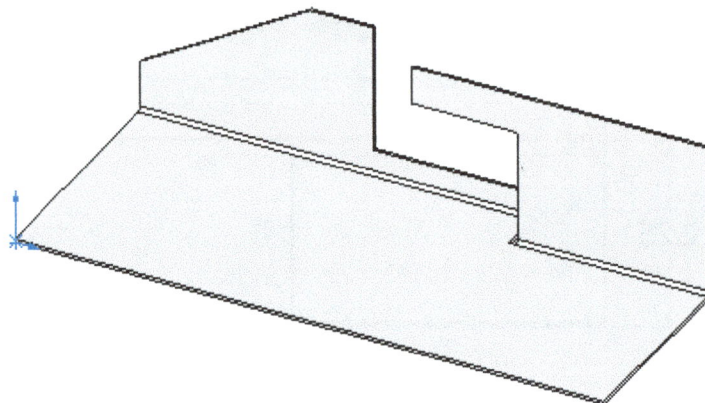

Add a Jog

Like in Chapter 4, select the **Jog** icon in the CommandManager, or pull down the "Insert" menu and select **Sheet Metal – Jog**.

When prompted in the PropertyManager to select a planer face on which to sketch bend lines, pick the front of the flange on the right of the part. When you select the planer face, a sketch opens on the plane.

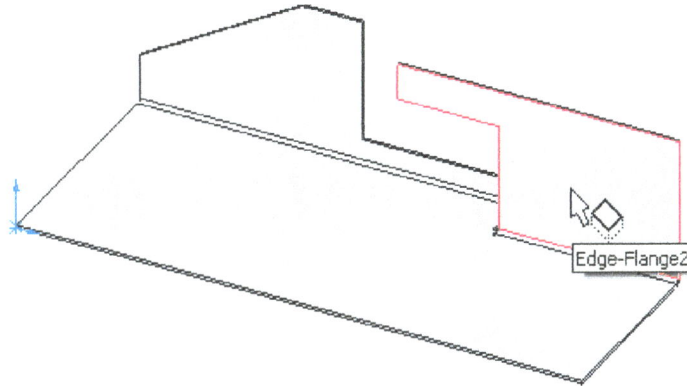

Select the **Line** icon in the CommandManager, or pull down the "Tools" menu and select **Sketch Entities – Line**.

Create a horizontal line on the selected surface as shown.

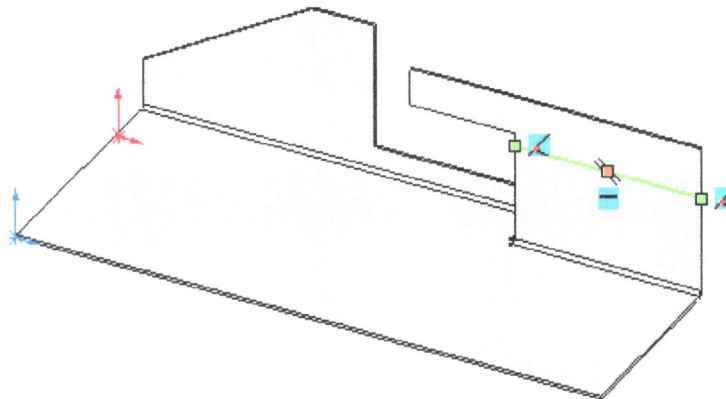

Select the **Smart Dimension** icon in the CommandManager, or pull down the "Tools" menu and select **Dimensions – Smart**.

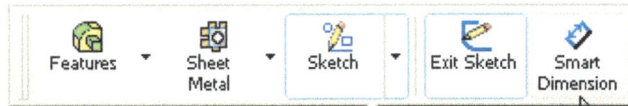

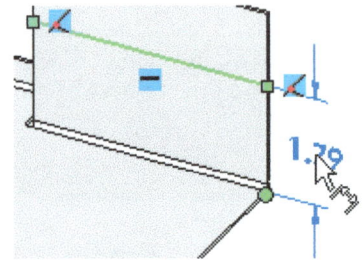

Place a '**2.00**' dimension from the sketched line to the bottom of the part bend line as shown.

Select the **Exit Sketch** icon in the CommandManager or in the upper right corner of the graphics area.

Select below the sketched horizontal line on the front of the flange for the fixed face. Remember that a large black dot will appear where you select on the part. Make sure the selection point is below the horizontal line.

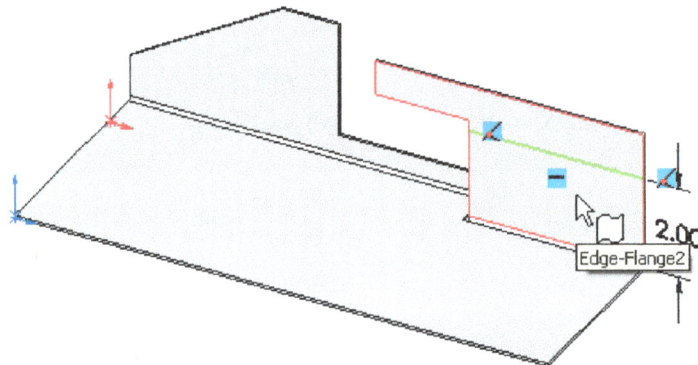

Set the **Jog Offset** to **Blind** distance of '**0.50in.**'

Make sure the **Outside Offset Dimension position** button is depressed and that the **Fix projected length** check box is checked.

Also, make sure that the **Material Inside Jog Position** button is depressed and that the **Jog Angle** is set to **90.00deg**.

Select the green check mark button at the top of the **Jog** PropertyManager to accept the settings and create the feature.

One More Flange

Select the **Edge Flange** icon in the CommandManager, or pull down the "Insert" menu and select **Sheet Metal – Edge Flange**.

Select the left most bottom edge as shown. Pull the preview flange to the right and click the left mouse button to set the flange in place.

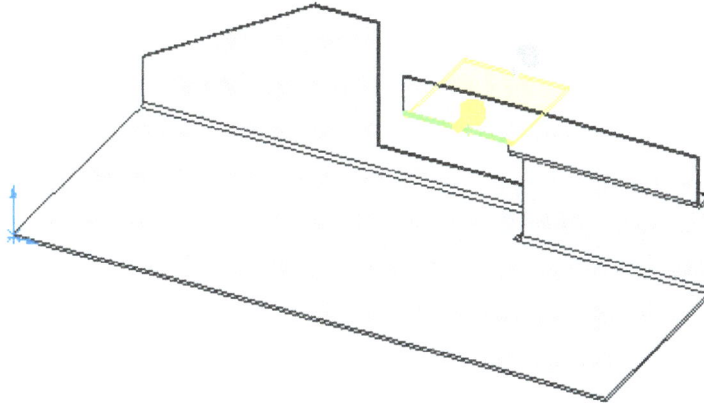

In the **Edge-Flange** PropertyManager, select the **Edit Flange Profile** button.

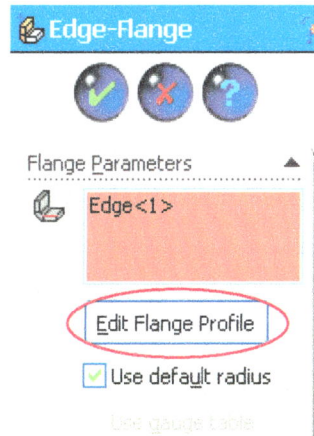

Change the display view to **Normal To** the sketch by selecting **Normal To** in the "View" pop-up menu in the lower left corner of the graphics area.

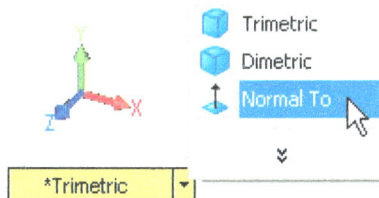

Left click and drag the right vertical line to the left.

Select the **Smart Dimension** icon in the CommandManager, or pull down the "Tools" menu and select **Dimensions – Smart**.

Place a '**0.25**' dimension from the right vertical sketched line to the left edge of the flange as shown.

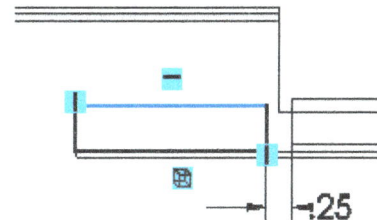

Select the **Back** button in the **Profile Sketch** dialog box.

Set the following settings: **Flange Length** to a **Blind** distance of '**0.50in**', depress the **Outer Virtual Sharp** button, depress the **Bend Outside Flange Position** button.

Select the green check mark button at the top of the **Edge-Flange** PropertyManager to accept the settings and create the feature.

Select the **Previous View** icon in the "View" toolbar, or pull down the "View" menu and select **Modify – Previous View** to return to the Trimetric view.

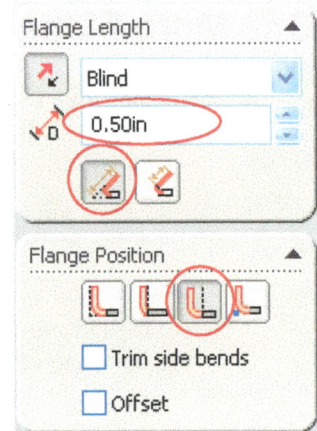

Previous View
Displays the previous view.

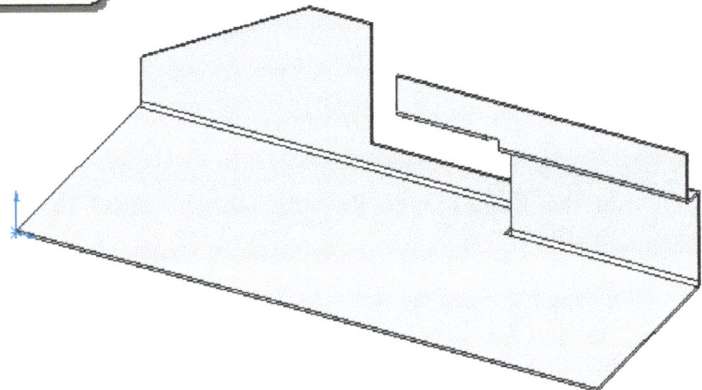

Add an Equation

The last edge flange should be at the same height as the top of the jog. To do this, in the FeatureManager design tree, right click on **Annotations** and select **Show Feature Dimensions**.

Double click on the '**2.00**' dimension. It should be on the right side of the part.

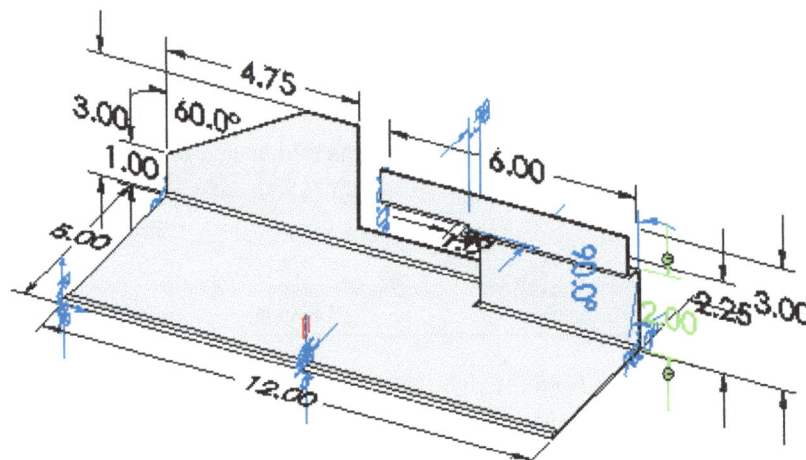

SolidWorks for the Sheet Metal Guy

Pull down the menu in the **Modify** dialog box and select **Add Equation**.

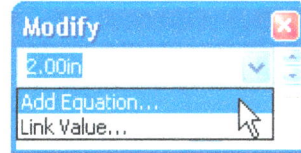

Select the '**2.25**' dimension in the graphics area. You may have to drag the dialog boxes out of the way in order to see the dimensions. Next, press the minus (-) button. Finally, select the '**R.08**' dimension in the graphics area.

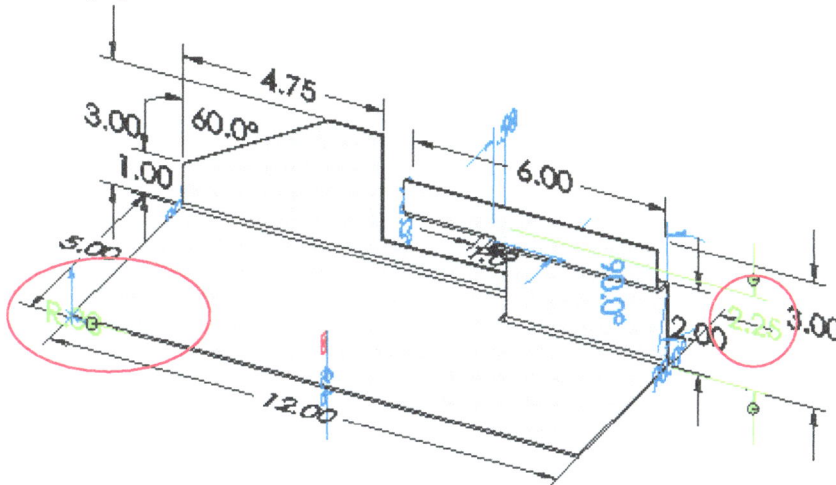

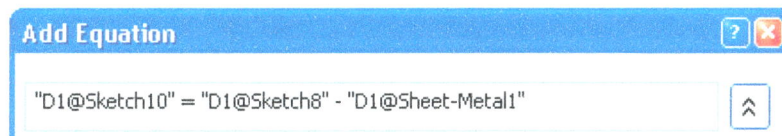

"D1@Sketch10" = "D1@Sketch8" - "D1@Sheet-Metal1"

Select the **OK** button in the **Add Equation** dialog box. Then, select the **OK** button in the **Equations** dialog box. The dimension will now look like below.

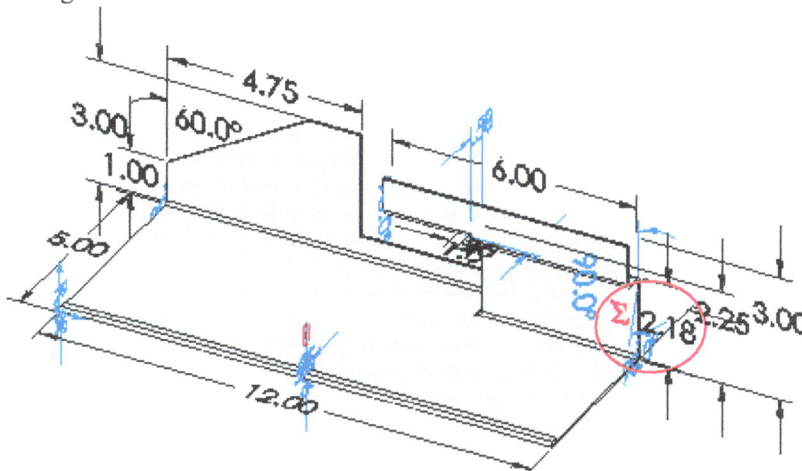

Select the **Rebuild** icon in the "Standard" toolbar to rebuild the part using the new dimension.

Add a Hem

Select the **Hem** icon in the CommandManager, or pull down the "Insert" menu and select **Sheet Metal – Hem**.

Hem
Curls edges of a sheet metal part.

Select the top front edge of the part as shown. If you select the lower front edge the hem will be below the part rather than on top of the part.

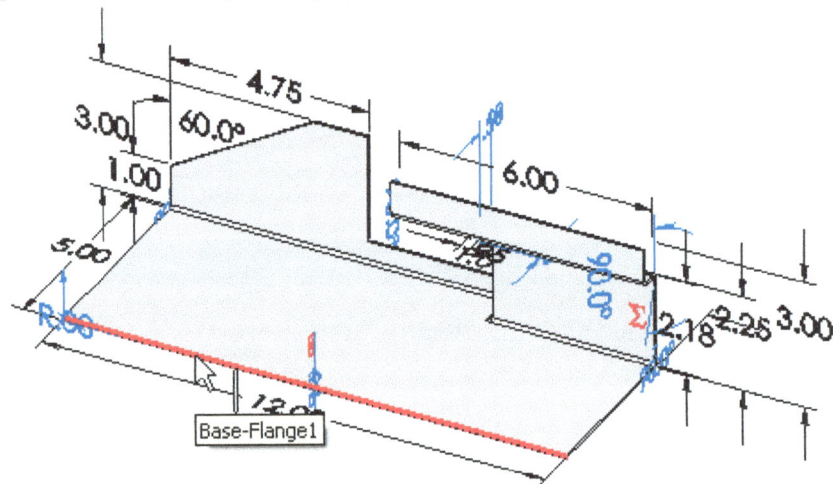

Select the **Material Inside** button.

Under **Type and Size**, select the **Closed** button and set the **Length** to '**0.25**.'

Select the green check mark button at the top of the **Hem** PropertyManager to accept the settings and create the feature.

Saving the Part

Select the **Save** icon in the "Standard" toolbar, or select **Save** from the "File" pull down menu.

The **Save As** dialog box appears.

In the **File name** box, type the name of the drawing number. For this lesson, use '**Bracket**' and select **Save**.

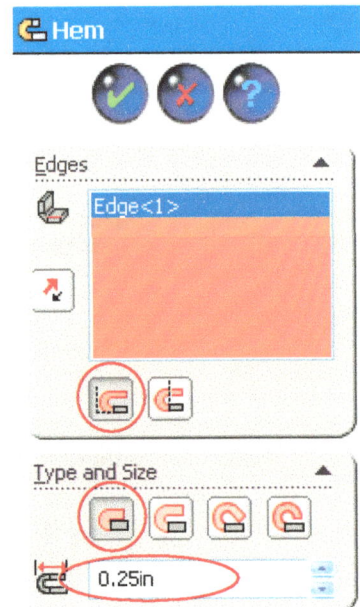

Chapter 6

Housing

The Housing provides slightly more complexity than the parts you have done so far. When adding flanges to the sides simultaneously, SolidWorks automatically miters the ends of the flanges to avoid interference.

A mirror is used to create two cutouts on the back flange. This makes the cutouts identical and when one is changed, the other will update.

The top of the part includes an offset bend, but the Jog command is not used to create it. An equation is used to maintain the offset distance equal to the material thickness.

If you want to see the effect of using the Jog command in this case, create the Jog before adding the side flanges.

Create a New Part Document

To begin this chapter, create a new part document by selecting the **New** icon in the "Standard" toolbar, or pull down the "File" menu and select **New**.

When the **New SolidWorks Document** dialog box appears, select **Part** and then select **OK**.

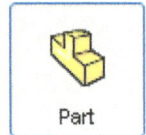

Create the Base Flange

Select the **Sheet Metal** icon in the control area of the CommandManager. Then, select the **Base-Flange/Tab** icon from the toolbar, or pull down the "Insert" menu and select **Sheet Metal – Base Flange**.

When you are prompted in the PropertyManager to select a plane on which to sketch the feature cross-section, select the **Top Plane** with the cursor.

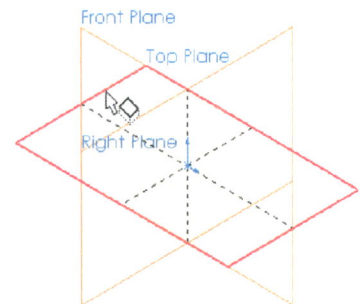

Create the Sketch

Select the **Rectangle** icon in the CommandManager, or pull down the "Tools" menu and select **Sketch Entities – Rectangle**.

Start the rectangle at the origin. Move the cursor up and to the right and create a rectangle.

Next, select the **Line** icon in the CommandManager, or pull down the "Tools" menu and select **Sketch Entities – Line**.

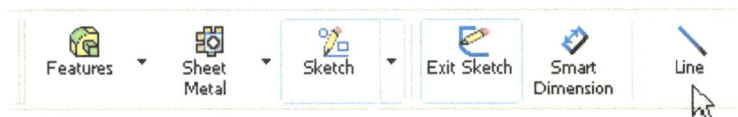

Create three new lines on the bottom line of the rectangle as shown.

Select the **Trim** icon in the CommandManager, or pull down the "Tools" menu and select **Sketch Tools – Trim**.

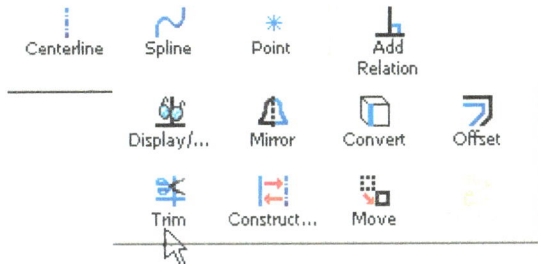

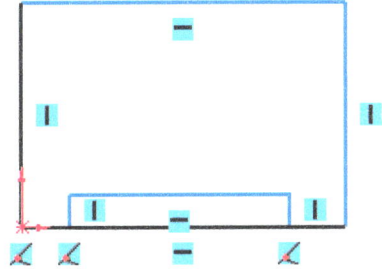

Select the middle of bottom horizontal line that you want to trim and click the left mouse button.

Select the **Smart Dimension** icon in the CommandManager, or pull down the "Tools" menu and select **Dimensions – Smart**, and dimension the sketch as shown.

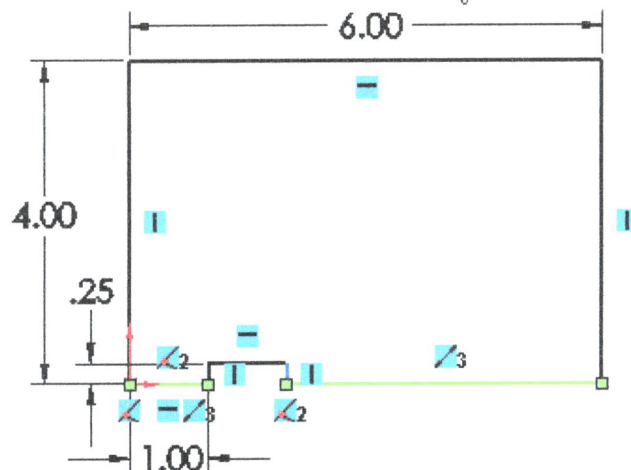

Select the **Add Relation** icon in the CommandManager, or pull down the "Tools" menu and select **Relations – Add**.

Select the bottom left horizontal line and the bottom right horizontal line as highlighted above.

In the **Add Relations** PropertyManager, select the **Equal** relation button.

Next, select the green check mark button at the top of the **Add Relations** PropertyManager.

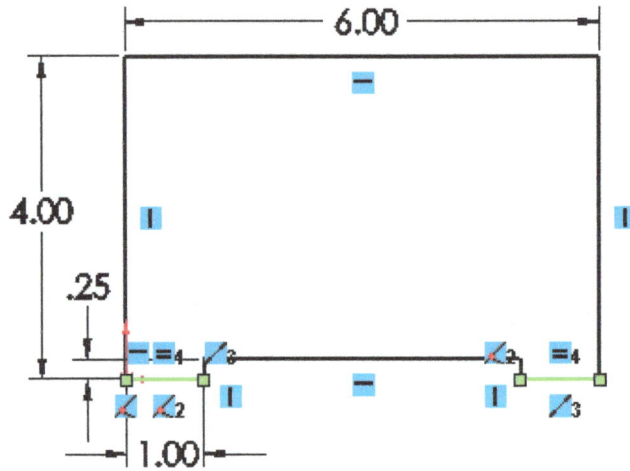

Select the **Exit Sketch** icon in the CommandManager or in the upper right corner of the graphics area.

Completing the Base Flange

In the **Base Flange** PropertyManager under **Sheet Metal Parameters**, set the **Thickness** to '**0.048in**'.

Check the **Reverse direction** check box to make sure that the preview of the extrusion is in the Y-positive direction (The yellow rectangle will be above the green rectangle).

Select the green check mark button at the top of the **Base Flange** PropertyManager to accept the settings and create the feature.

In the FeatureManager design tree, right click on the **Sheet-Metal1** item and select **Edit Feature** from the menu.

In the **Sheet-Metal1** Property-Manager, enter '**0.05in**' for the default **Bend Radius** and then select the green check mark button.

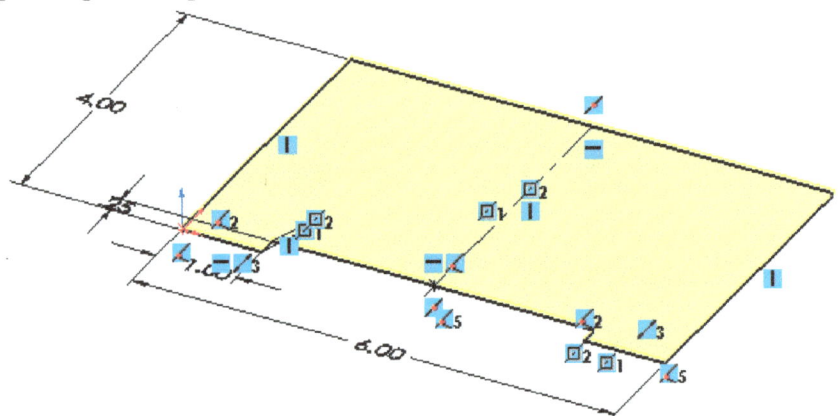

Create the Front Flanges

Select the **Edge Flange** icon in the CommandManager, or pull down the "Insert" menu and select **Sheet Metal – Edge Flange**.

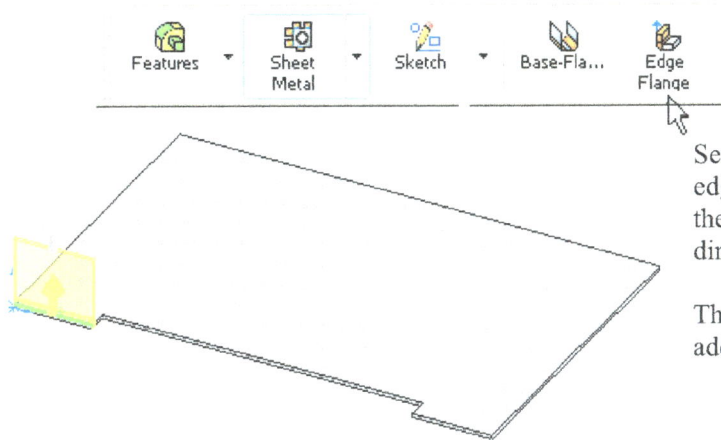

Select the front left edge as the edge to add the flange to. Move the cursor up and click to set the direction of the flange.

Then, select the front right edge to add a second flange.

In the **Edge-Flange** PropertyManager, set the **Flange Length** to '**0.50in**', depress the **Outer Virtual Sharp** icon, and depress the **Material Inside Flange Position** icon.

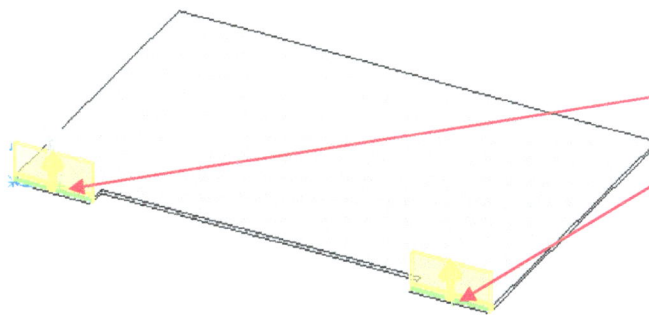

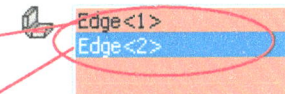

With **Edge<2>** highlighted in the **Edge-Flange** PropertyManager, select the **Edit Flange Profile** button.

Left click and drag the left side of the profile to the right as shown.

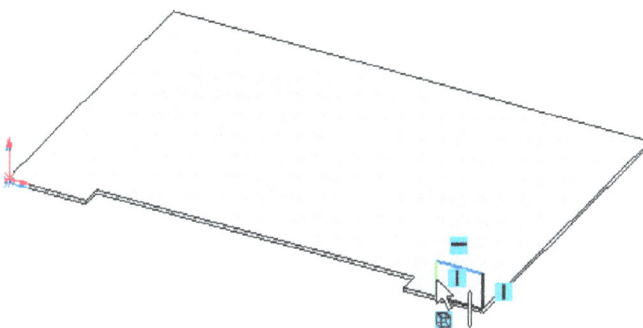

Select the **Smart Dimension** icon in the CommandManager, or pull down the "Tools" menu and select **Dimensions – Smart**.

Add a '**0.50**' dimension to the bottom line of the sketch as shown below.

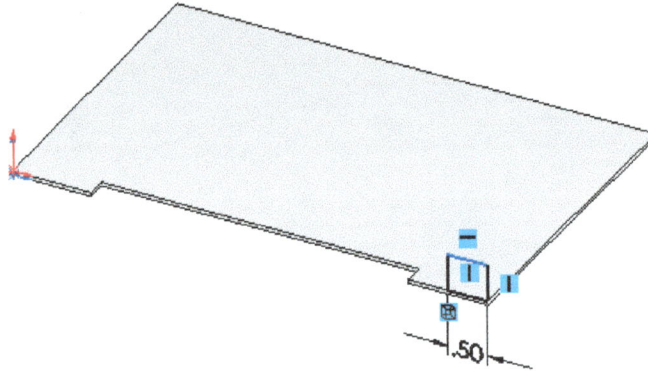

Select the **Back** button in the **Profile Sketch** dialog box.

Left click on **Edge<1>** in the **Edge-Flange** PropertyManager to highlight it, and then select the **Edit Flange Profile** button.

The left profile sketch will become active. Left click and drag the right side of the profile to the left. Then, use **Smart Dimension** to add a '**0.50**' dimension to the bottom line of the sketch as shown.

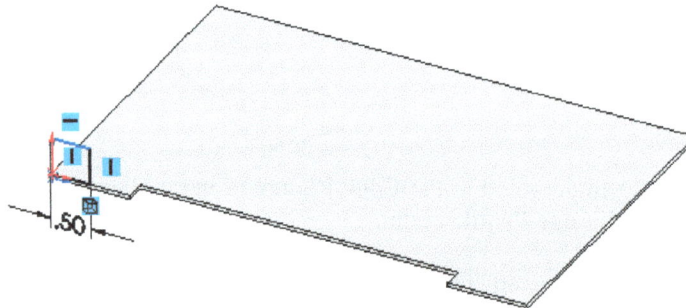

Select the **Back** button in the **Profile Sketch** dialog box.

Select the green check mark button at the top of the **Edge-Flange** PropertyManager to accept the settings and create the feature.

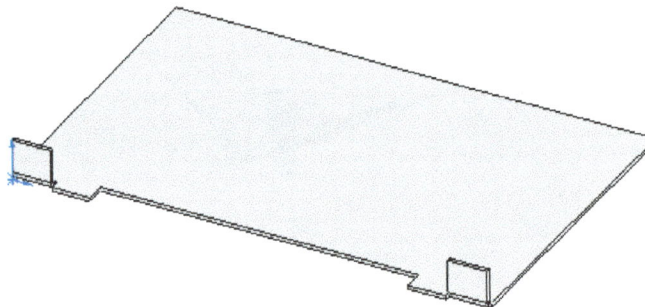

Create the Center Flange

Select the **Edge Flange** icon in the CommandManager, or pull down the "Insert" menu and select **Sheet Metal – Edge Flange**.

Select the front edge of the cutout. Move the cursor up and click to set the direction of the flange.

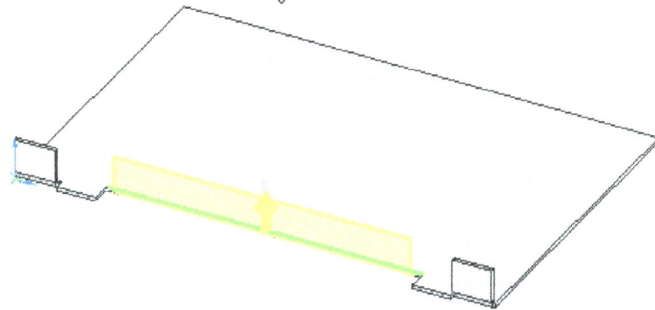

In the **Edge-Flange** PropertyManager, set the **Flange Length** to '**0.50in**' and depress the **Outer Virtual Sharp** button like above.

Select the **Bend Outside Flange Position** button.

Then, select the **Edit Flange Profile** button.

Left click and drag the right vertical line to the left and the left vertical line to the right as shown. Again, the actual distance does not matter, as this will be controlled by dimensions.

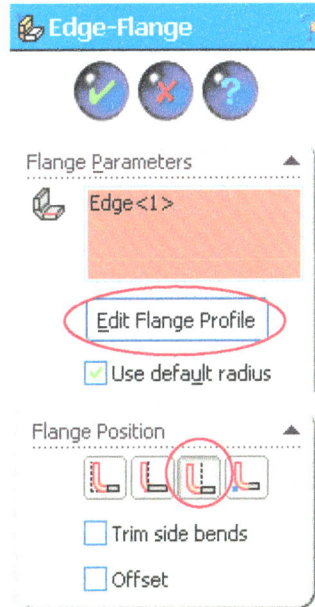

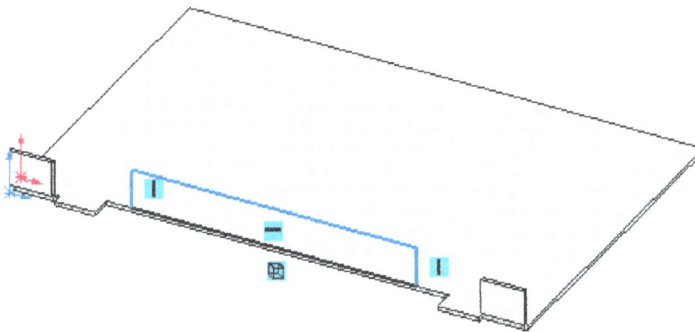

Select the **Smart Dimension** icon in the CommandManager, or pull down the "Tools" menu and select **Dimensions – Smart**.

Add a '**0.125**' dimension to each side from the vertical line to the corner point as shown.

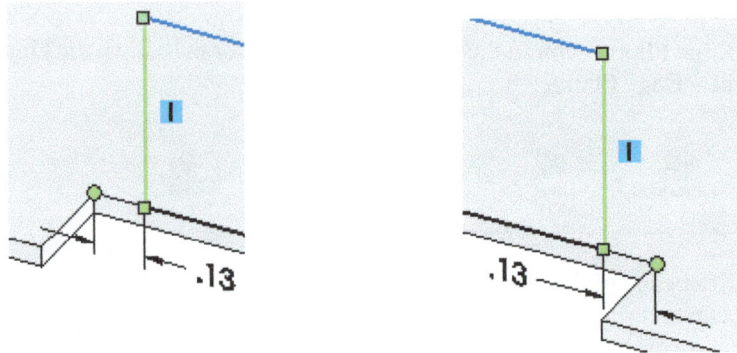

Select the **Back** button in the **Profile Sketch** dialog box.

Select the green check mark button at the top of the **Edge-Flange** PropertyManager to accept the settings and create the center flange.

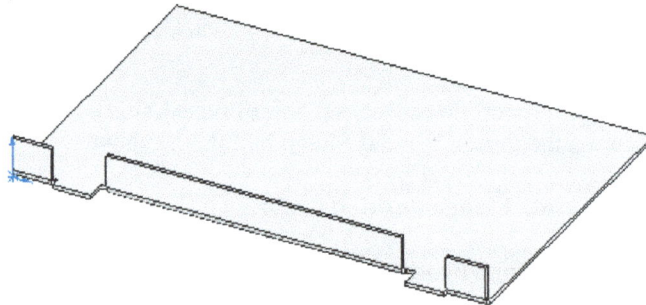

Create the Back Flange

Select the **Edge Flange** icon in the CommandManager, or pull down the "Insert" menu and select **Sheet Metal – Edge Flange**.

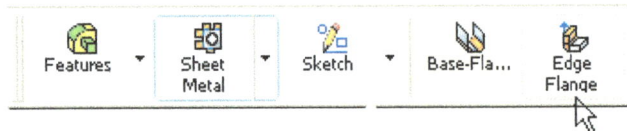

Select the back edge of the part. Move the cursor up and click to set the direction of the flange.

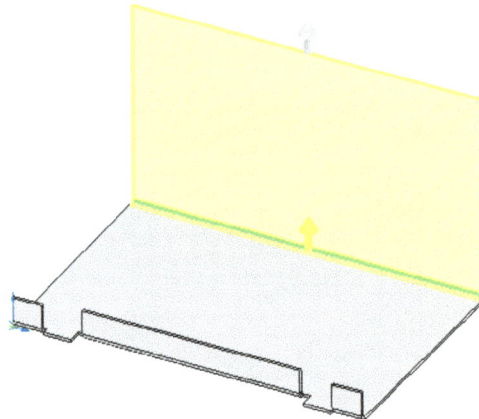

In the **Edge-Flange** PropertyManager, set the **Flange Length** to '5.00in', depress the **Outer Virtual Sharp** button, and depress the **Material Inside** button.

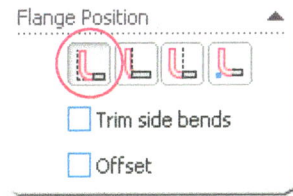

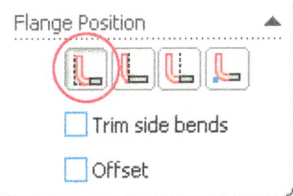

Select the green check mark button at the top of the **Edge-Flange** PropertyManager to accept the settings and create the feature.

Creating the Top Flange

Select the **Edge Flange** icon in the CommandManager, or pull down the "Insert" menu and select **Sheet Metal – Edge Flange**.

Select the top edge of the back flange. Move the cursor to the left and click to set the direction of the flange.

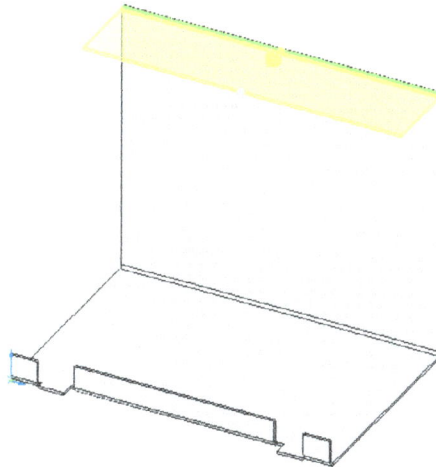

In the **Edge-Flange** PropertyManager, set the **Flange Length** to '1.50in', depress the **Outer Virtual Sharp** button, and depress the **Material Inside** button.

Select the green check mark button at the top of the **Edge-Flange** PropertyManager to accept the settings and create the top flange.

Creating the Side Flanges

Select the **Edge Flange** icon in the CommandManager, or pull down the "Insert" menu and select **Sheet Metal – Edge Flange**.

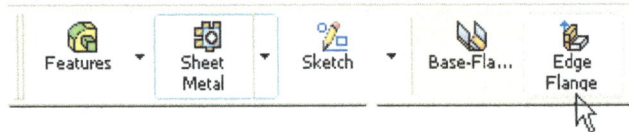

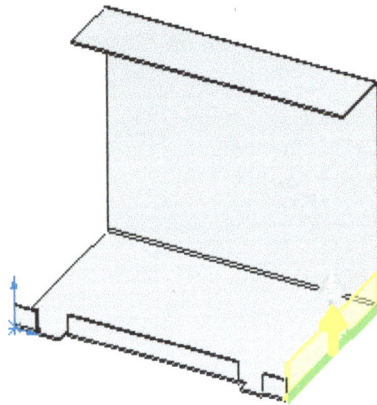

Select the right most bottom edge as shown. Pull the preview flange up and click the left mouse button to set the flange in place.

Select the other five side edge lines to create a total of six edge flanges.

In the **Edge-Flange** PropertyManager, set the **Flange Length** to '**0.25in**', depress the **Outer Virtual Sharp** button, and depress the **Material Inside** button. Check the **Trim side bends** check box.

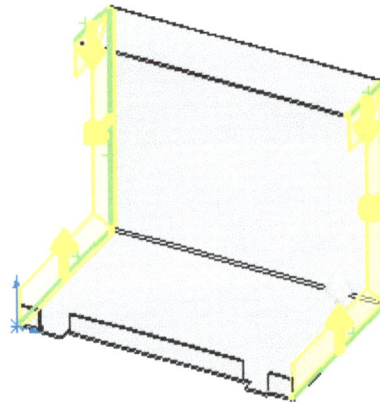

Select the green check mark button at the top of the **Edge-Flange** PropertyManager to accept the settings and create the side flanges.

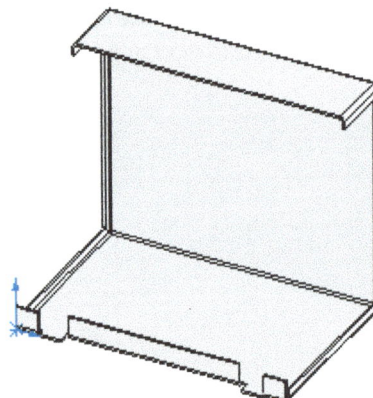

Add a Jog to the Top Flange

While the goal is to offset the flange by a material thickness, you will use the Miter Flange command. The Jog command doesn't create the corners in the desired manner relative to the Edge (side) Flanges.

Start by creating a plane onto which you will sketch the flange profile. Pull down the "Insert" menu and select **Reference Geometry – Plane**.

In the upper left hand corner of the graphics area, click on the plus sign next to the part name (Part1) to expand the design tree.

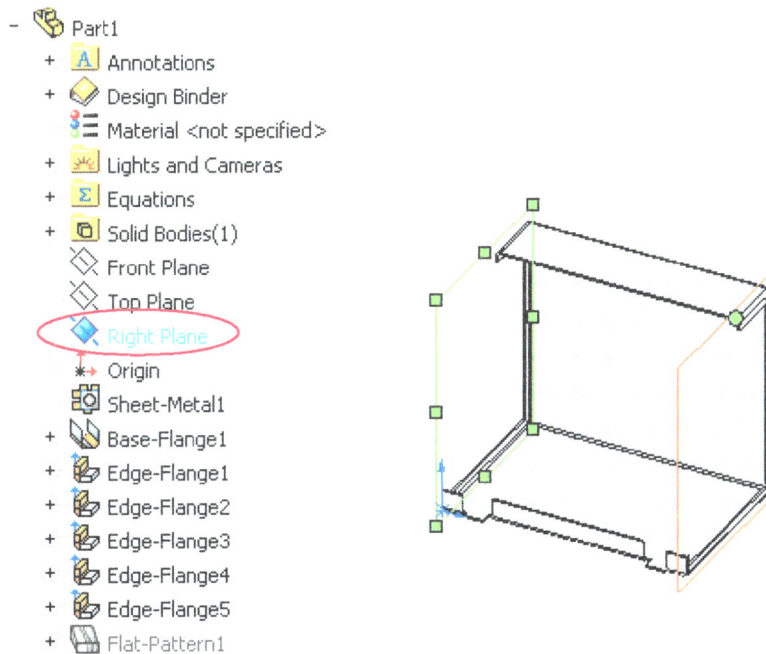

Left click on **Right Plane**. The Right Plane will highlight in the graphics area when you place the cursor over the **Right Plane** in the design tree.

In the **Plane** PropertyManager, select the **Parallel Plane at Point** button.

In the graphics area, select one of the points of the vertical right edge line where the top flange meets the bend area of the side edge flange.

A parallel plane is created for you at the selected point.

Select the green check mark button at the top of the **Plane** PropertyManager to create the plane.

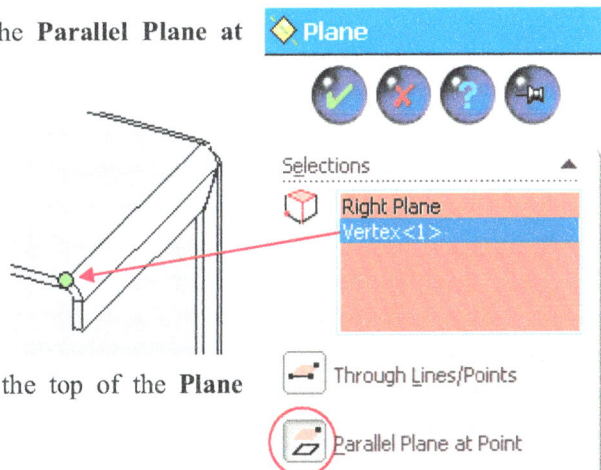

Select the **Miter Flange** icon in the CommandManager, or pull down the "Insert" menu and select **Sheet Metal – Miter Flange**.

You will be sketching the profile of the jog. If the plane you just created is not selected, you are prompted in the PropertyManager to select a plane, a planar face or an edge on which to sketch the feature cross-section. Select **Plane1** in the graphics area.

A sketch is started for you. Change the display view to **Normal To** the sketch by selecting **Normal To** in the View list in the lower left corner of the graphics area.

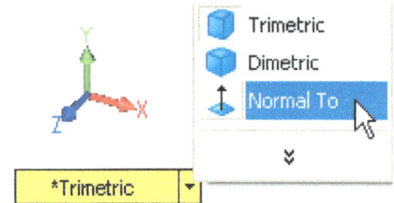

Select the **Line** icon in the CommandManager, or pull down the "Tools" menu and select **Sketch Entities – Line**.

Use the **Line** tool to create the side profile of the flange as shown.

Start by creating a diagonal line, beginning at the upper left corner of the flange as shown, and go up and to the left and create the line.

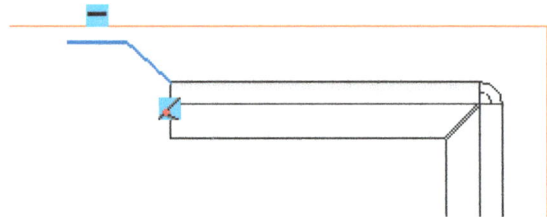

Then, create a short horizontal line to the left as shown.

Select the **Smart Dimension** icon in the CommandManager, or pull down the "Tools" menu and select **Dimensions – Smart**.

Dimension the sketch as shown. Create the horizontal and angular dimensions first. Then, add a vertical dimension between the horizontal line and the top edge line of the part.

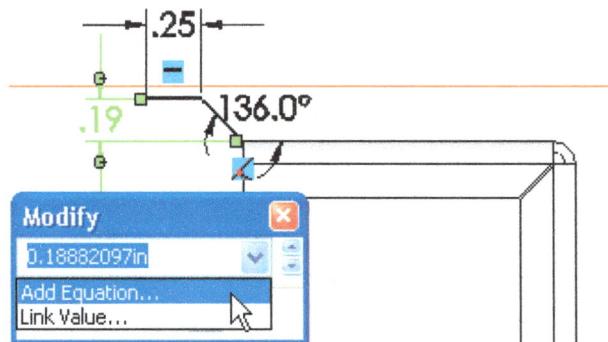

In the **Modify** dialog box, pull down the menu and select **Add Equation**.

The **Add Equation** dialog box appears and the name of the dimension is already entered for you.

Type '**Thickness**' and then select the **OK** button.

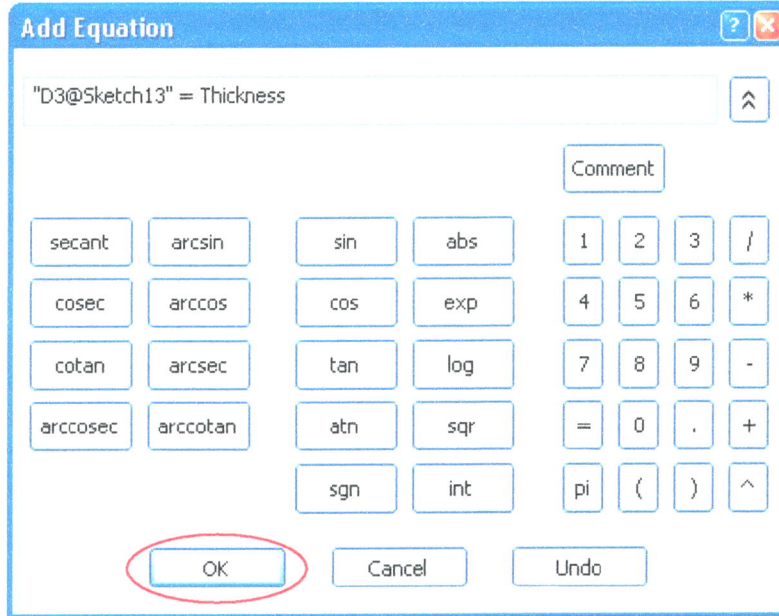

Select the **OK** button again to close the **Equations** dialog box.

The dimension is now updated, indicating that it is an equation.

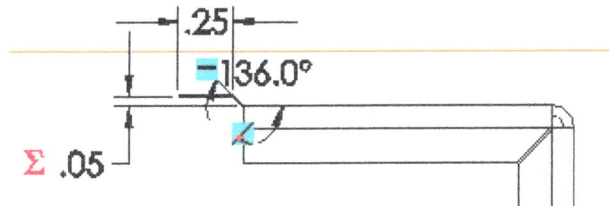

Select the **Exit Sketch** icon in the CommandManager or in the upper right corner of the graphics area.

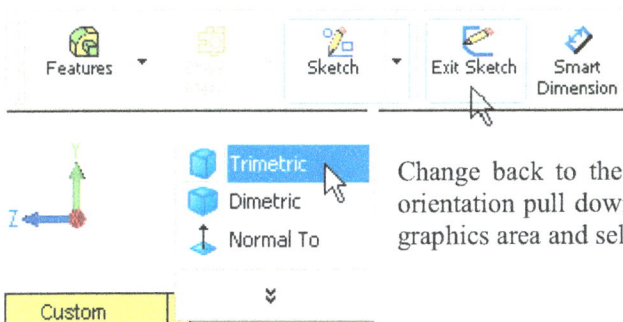

Change back to the **Trimetric** view by selecting the View orientation pull down arrow in the bottom left corner of your graphics area and selecting **Trimetric**.

A preview of the flange will be shown.

In the **Miter Flange** PropertyManager, make sure that the **Flange position** is set to **Material Outside**.

Select the green check mark button at the top of the **Miter Flange** PropertyManager to accept the settings and create the feature.

In the FeatureManager design tree, right click on **Plane1** and select **Hide**.

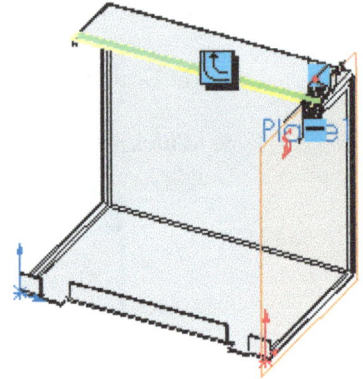

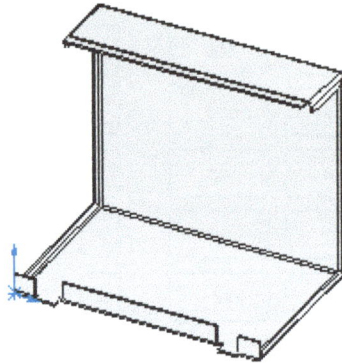

Adding a Cutout on the Top Flange

Select the **Extruded Cut** icon in the CommandManager, or pull down the "Insert" menu and select **Cut – Extrude**.

Select the top of the part on the top flange for the sketch plane.

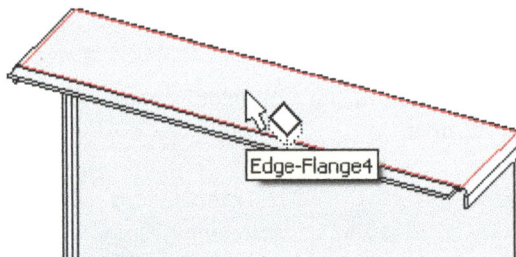

Change the display view to **Normal To** the sketch by selecting **Normal To** in the View list in the lower left corner of the graphics area.

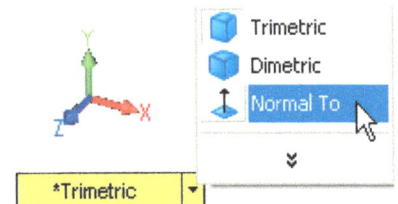

SolidWorks for the Sheet Metal Guy

Select the **Rectangle** icon in the CommandManager, or pull down the "Tools" menu and select **Sketch Entities – Rectangle**.

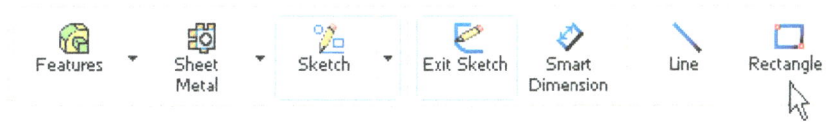

Create a rectangle on the top face as shown.

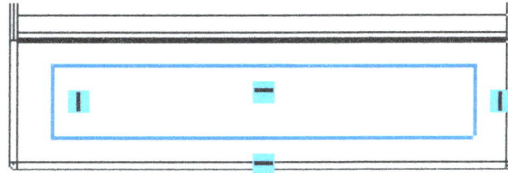

Select the **Smart Dimension** icon in the CommandManager, or pull down the "Tools" menu and select **Dimensions – Smart**.

Dimension the sketch as shown, adding the '**0.50**' dimensions.

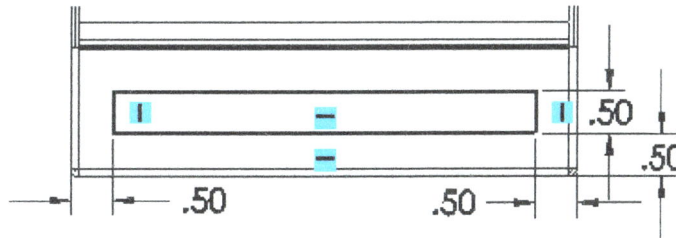

Select the **Exit Sketch** icon in the CommandManager or in the upper right corner of the graphics area.

In the **Cut-Extrude** Property Manager, check the **Link to thickness** check box.

Select the green check mark button at the top of the **Cut-Extrude** PropertyManager to accept the settings and create the feature.

Change the display view to **Trimetric** by selecting **Trimetric** in the View list in the lower left corner of the graphics area.

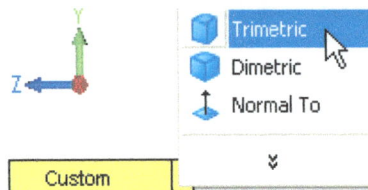

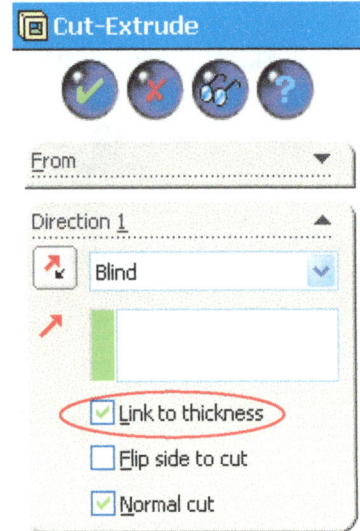

Adding Cutouts on the Back Flange

Select the **Extruded Cut** icon in the CommandManager, or pull down the "Insert" menu and select **Cut – Extrude**.

Select the back flange for the sketch plane.

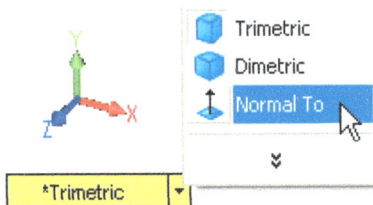

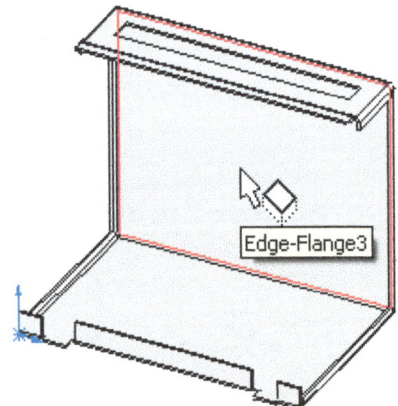

Change the display view to **Normal To** the sketch by selecting **Normal To** in the View list in the lower left corner of the graphics area.

Select the **Rectangle** icon in the CommandManager, or pull down the "Tools" menu and select **Sketch Entities – Rectangle**.

Create a rectangle on the left side of the back flange.

Select the **Smart Dimension** icon in the CommandManager, or pull down the "Tools" menu and select **Dimensions – Smart**, and dimension the sketch as shown.

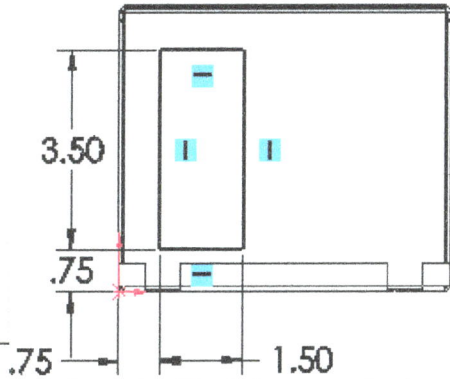

Select the **Centerline** icon in the CommandManager, or pull down the "Tools" menu and select **Sketch Entities – Centerline**, and create a vertical centerline connecting the two midpoints of the top and bottom horizontal lines of the part as shown below.

Then, select the **Mirror** icon in the CommandManager, or pull down the "Tools" menu and select **Sketch Tools – Mirror**.

Select the four lines you created which form the rectangle as the entities to mirror.

Left click inside the **Mirror about** box and then select the centerline. You may need to select the centerline twice to deselect and then select it as the line to mirror about.

The sketch is mirrored to the other side of the centerline.

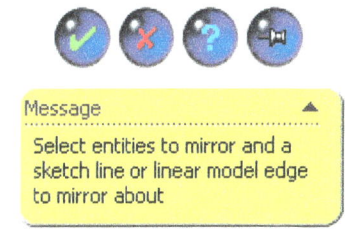

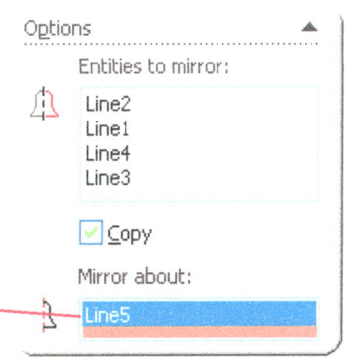

Select the green check mark button at the top of the **Mirror** PropertyManager.

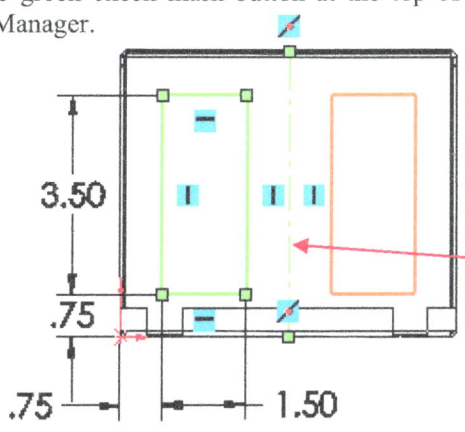

Select the **Exit Sketch** icon in the CommandManager or in the upper right corner of the graphics area.

In the **Cut-Extrude** PropertyManager, check the **Link to thickness** check box.

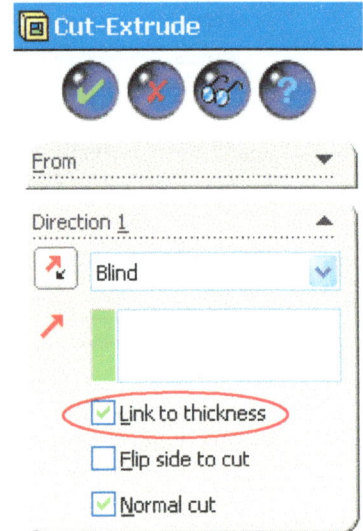

Select the green check mark button at the top of the **Cut-Extrude** PropertyManager to accept the settings and create the feature.

Change the display view to **Trimetric** by selecting **Trimetric** in the View list in the lower left corner of the graphics area.

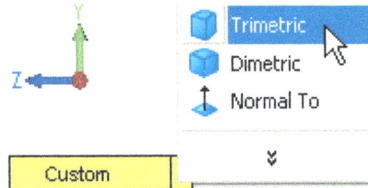

Saving the Part

Select the **Save** icon in the "Standard" toolbar, or select **Save** from the "File" pull down menu.

The **Save As** dialog box appears.

In the **File name** box, type the name of the drawing number. For this chapter, use '**10-3-4332-5**' and select **Save**.

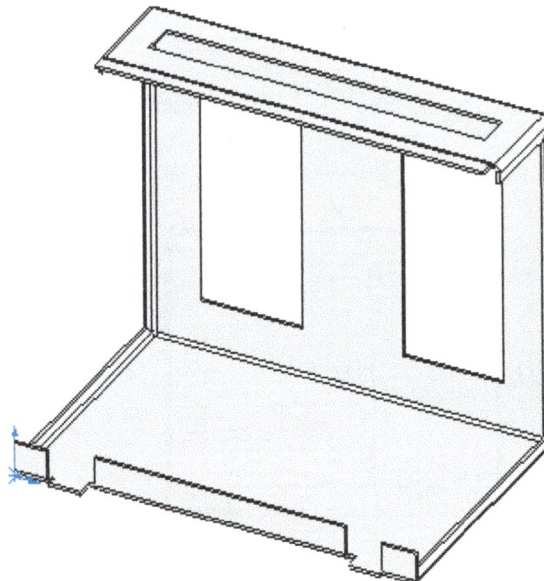

Chapter 7

Mounting Bracket

The Mounting Bracket shows the use of the Base Flange/Tab command to add more material to the part after the Jog command. This was done primarily to show that you can do this. It is actually easier to create the part by editing the original Base Flange sketch to include the additional feature and then add the jog to the part.

Two different methods of filleting the corners are then used because of this.

The Mirror command is used to copy the tab, ensuring that they will always be identical. These tabs, if modified, may overlap each other in the flat. After creating the part, you will unfold it to check for this conflict.

Create a New Part Document

Select the **New** icon in the "Standard" toolbar, or pull down the "File" menu and select **New**.

When the **New SolidWorks Document** dialog box appears, select **Part** and then select **OK**.

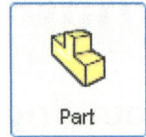

To increase your graphics area size, you can reduce the amount of space the CommandManager takes up by turning off the icon descriptions.

To do this, right click in the control area of the CommandManager and select **Show Description**. This will uncheck **Show Description** and the CommandManager will look like the other toolbars.

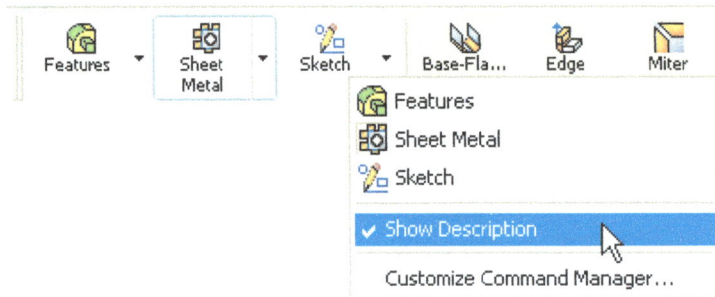

Creating the Base Flange

Select the **Sheet Metal** icon in the control area of the CommandManager. Select the **Base-Flange/Tab** icon from the toolbar, or pull down the "Insert" menu and select **Sheet Metal – Base Flange**.

When you are prompted in the PropertyManager to select a plane on which to sketch the feature cross-section, select the **Top Plane** with the cursor.

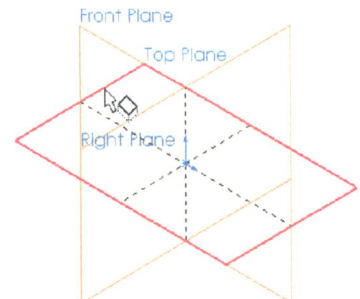

Select the **Rectangle** icon in the CommandManager, or pull down the "Tools" menu and select **Sketch Entities – Rectangle**, and start the rectangle at the origin, and move the cursor up and to the right and create a rectangle.

Next, select the **Line** icon in the CommandManager, or pull down the "Tools" menu and select **Sketch Entities – Line**.

Create a small horizontal line starting on the right vertical line. Then, create a small vertical line starting at the endpoint of the new horizontal line. Finally, create another horizontal line starting at the endpoint of the new vertical line and ending on the right most vertical line.

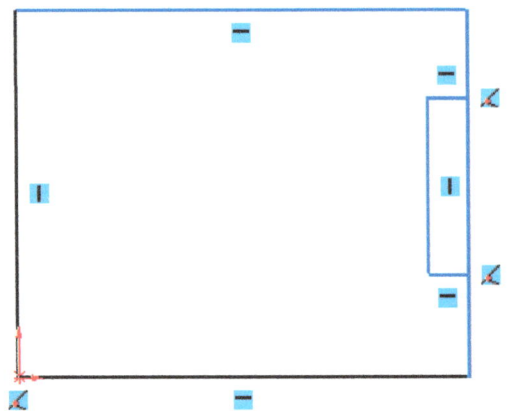

Select the **Trim** icon in the CommandManager, or pull down the "Tools" menu and select **Sketch Tools – Trim**.

Select the middle of right vertical line that you want to trim and press the left mouse button.

Dimension the Sketch

Select the **Smart Dimension** icon in the CommandManager, or pull down the "Tools" menu and select **Dimensions – Smart**. Dimension the sketch as shown.

Select the **Exit Sketch** icon in the CommandManager or in the upper right corner of the graphics area.

Completing the Base Flange

In the **Base Flange** PropertyManager under **Sheet Metal Parameters**, set the **Thickness** to **0.06**.

Check the **Reverse direction** check box to make sure that the preview of the extrusion is in the Y-positive direction (The yellow rectangle will be above the green rectangle).

Select the green check mark button at the top of the **Base Flange** PropertyManager to accept the settings and create the feature.

Create a Tab

Select the **Sheet Metal** icon in the control area of the CommandManager. Then, select the **Edge Flange** icon in the CommandManager, or pull down the "Insert" menu and select **Sheet Metal – Edge Flange**.

Select the front right edge, and move the cursor up and click to set the direction of the flange.

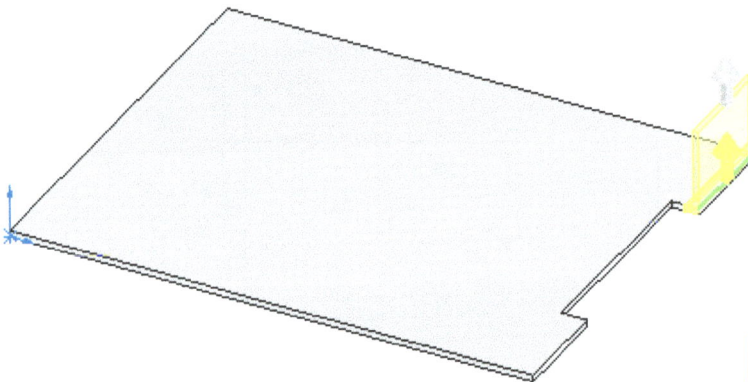

In the **Edge-Flange** PropertyManager, select the **Edit Flange Profile** button.

Change the display view to **Normal To** the sketch by selecting **Normal To** in the View list in the lower left corner of the graphics area.

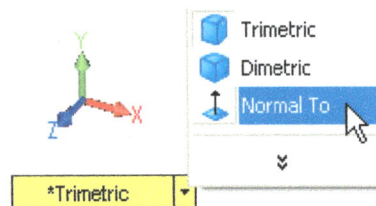

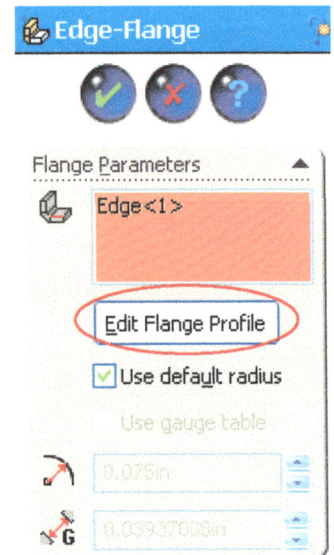

SolidWorks for the Sheet Metal Guy

Select the **Line** icon in the CommandManager, or pull down the "Tools" menu and select **Sketch Entities – Line**.

Create two parallel angled lines. Use the inferencing lines to make sure that the lines are parallel and that the endpoints are perpendicular as shown.

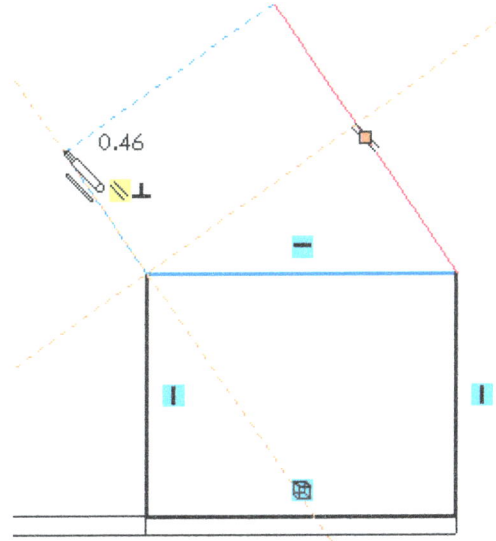

Then, create a line connecting the endpoints of the two angled lines.

Press the **Escape** key on the keyboard to deactivate the **Line** tool. Then, left click on the center horizontal line and press the **Delete** key on the keyboard.

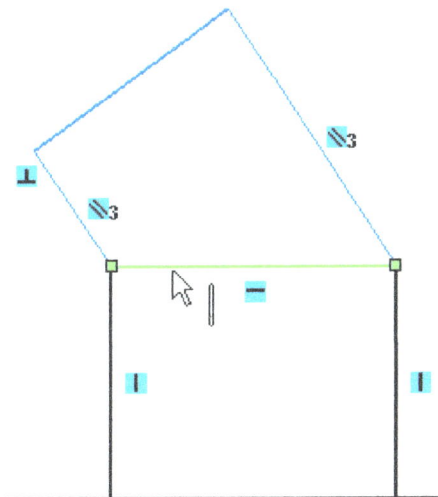

Select the **Sketch Fillet** icon in the CommandManager, or pull down the "Tools" menu and select **Sketch Tools – Fillet**.

Sketch Fillet
Rounds the corner at the intersection of two sketch entities, creating a tangent arc.

In the **Sketch Fillet** PropertyManager, set the **Radius** to '**0.25in**'.

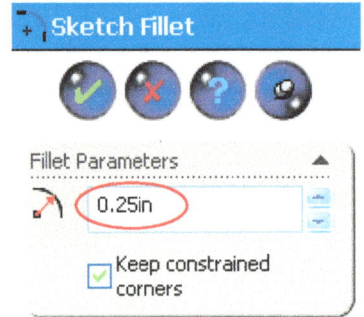

Select the intersection points of the vertical lines and the angled lines to add two fillets.

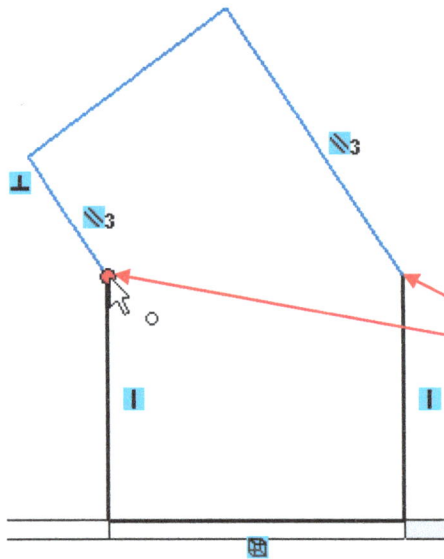

Sketch Fillet

Fillet Parameters

0.25in

☑ Keep constrained corners

Select these intersection points

Select the **Smart Dimension** icon in the CommandManager, or pull down the "Tools" menu and select **Dimensions – Smart**.

Dimension the sketch as shown.

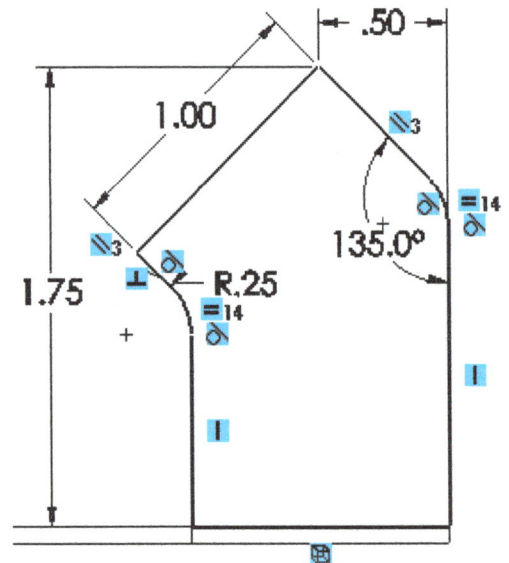

.50

1.00

135.0°

1.75

R.25

Select the **Back** button in the **Profile Sketch** dialog box. < Back

In the **Edge-Flange** PropertyManager, in the **Flange Position** window, select the **Material Inside Flange Position** button.

Select the green check mark button at the top of the **Edge-Flange** PropertyManager to accept the settings and create the feature.

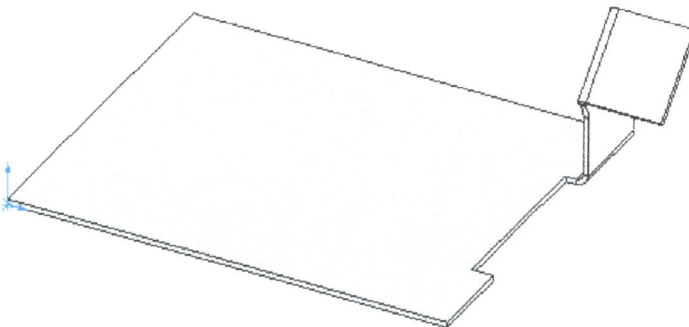

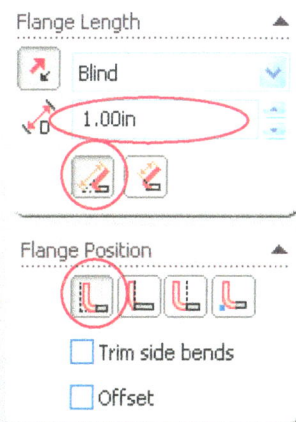

Select the **Previous View** icon in the "View" toolbar, or pull down the "View" menu and select **Modify – Previous View** to return to the Trimetric view.

Add a Flange to the Tab

Select the **Edge Flange** icon in the CommandManager, or pull down the "Insert" menu and select **Sheet Metal – Edge Flange**.

Select the front edge of the flange and move the cursor to the right and click to set the direction of the flange.

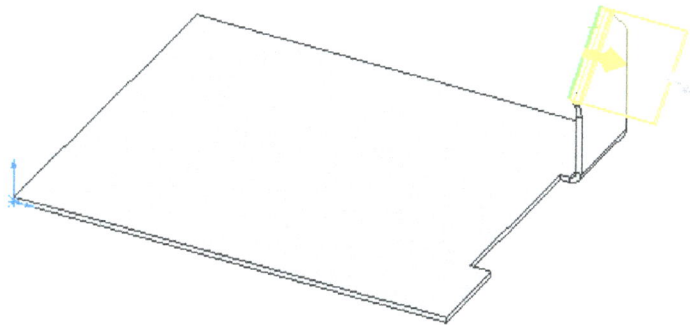

In the **Edge-Flange** PropertyManager, set the **Flange Length** to '**1.00in**', depress the **Outer Virtual Sharp** button, and depress the **Material Inside Flange Position** button.

Select the green check mark button at the top of the **Edge-Flange** PropertyManager to accept the settings and create the feature.

Add Another Flange to the Tab

Select the **Edge Flange** icon in the CommandManager, or pull down the "Insert" menu and select **Sheet Metal – Edge Flange**.

Select the front edge of the flange and move the cursor down and to the right and click to set the direction of the flange.

In the **Edge-Flange** PropertyManager, select the **Edit Flange Profile** button.

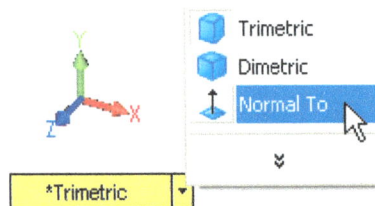

Change the display view to **Normal To** the sketch by selecting **Normal To** in the "View" pop-up menu in the lower left corner of the graphics area.

Select the **Smart Dimension** icon in the CommandManager, or pull down the "Tools" menu and select **Dimensions – Smart**.

Add a '**0.375**' dimension to the right vertical line.

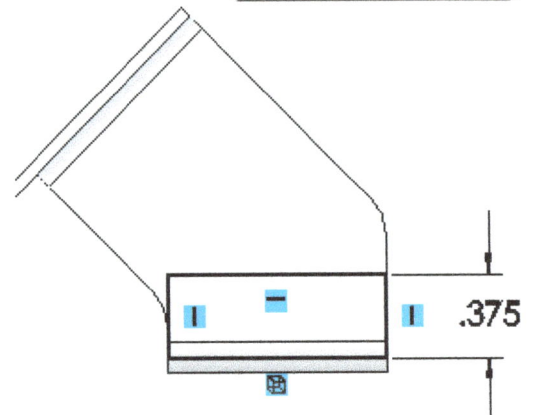

Select the **Sketch Fillet** icon in the CommandManager, or pull down the "Tools" menu and select **Sketch Tools – Fillet**.

In the **Sketch Fillet** PropertyManager, set the **Radius** to '**0.25in**.'

Select the upper right and upper left corners, adding two fillets.

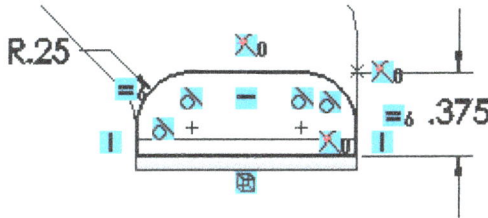

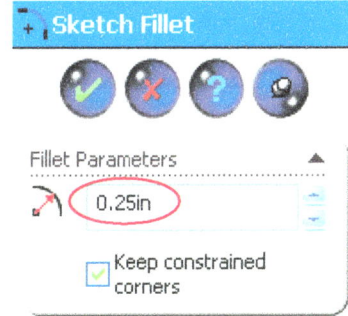

Select the **Back** button in the **Profile Sketch** dialog box.

In the **Edge-Flange** PropertyManager, select the **Material Inside** button under **Flange Position**.

Select the green check mark button at the top of the **Edge-Flange** PropertyManager to accept the settings and create the flange.

Select the **Previous View** icon in the "View" toolbar, or pull down the "View" menu and select **Modify – Previous View** to return to the Trimetric view.

Your part should now look like this.

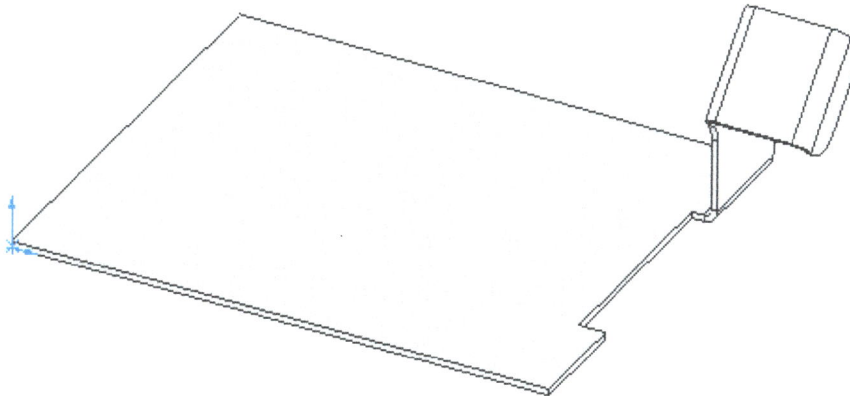

Mirroring the Tab

SolidWorks allows you to create a copy of a feature or multiple features by mirroring about a face or a plane. First, you must create a plane through the center of the part.

To do this, pull down the "Insert" menu, and select **Reference Geometry – Plane**.

In the upper left hand corner of the graphics area, click on the plus sign next to the document name (Part1) to expand the flyout FeatureManager design tree.

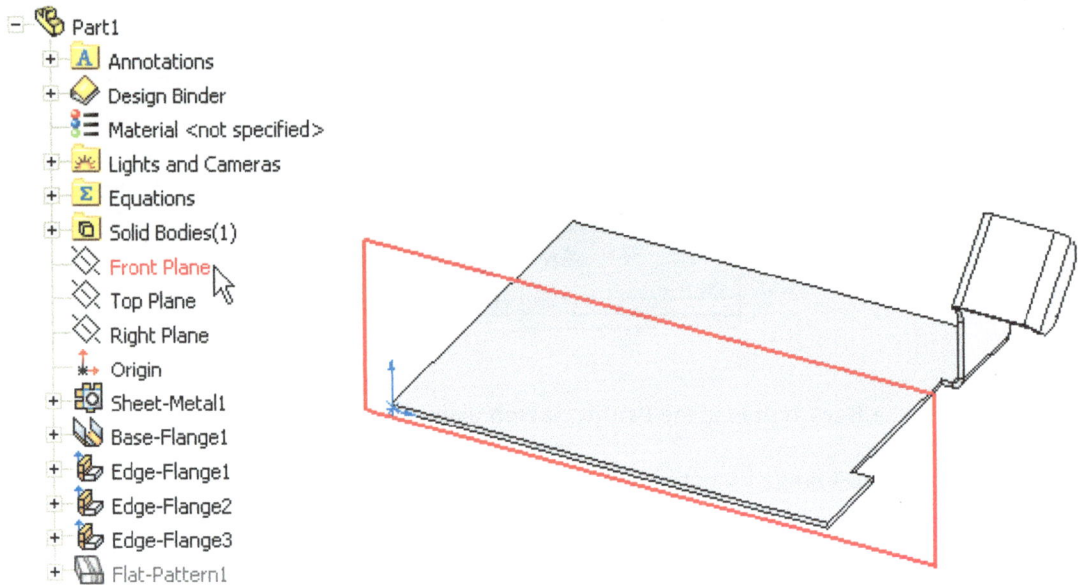

Left click on **Front Plane**. The Front Plane will highlight in the graphics area when you place the cursor over the **Front Plane** in the flyout FeatureManager design tree.

In the **Plane** PropertyManager, select the **Parallel Plane at Point** button.

In the graphics area, select the midpoint of the right edge line of the part inside of the notch.

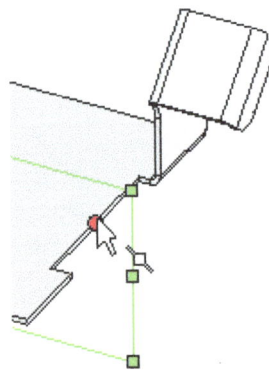

A parallel plane is created for you at the selected point.

Select the green check mark button at the top of the **Plane** PropertyManager.

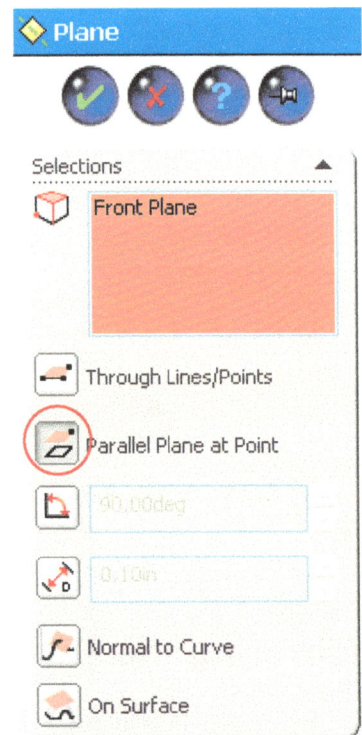

Select the **Features** icon in the control area of the CommandManager. Then, select the **Mirror** icon from the toolbar, or pull down the "Insert" menu and select **Pattern/Mirror – Mirror**.

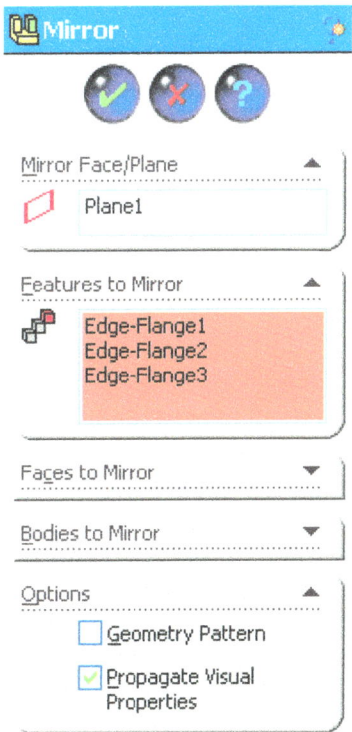

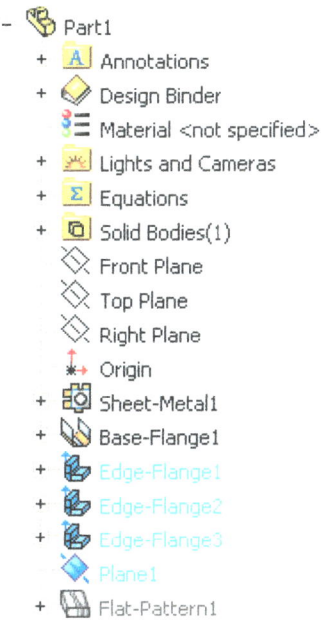

Features
Feature commands

Mirror
Mirrors features, faces, and bodies about a face or a plane.

Mirror

Mirror Face/Plane
Plane1

Features to Mirror
Edge-Flange1
Edge-Flange2
Edge-Flange3

Faces to Mirror

Bodies to Mirror

Options
☐ Geometry Pattern
☑ Propagate Visual Properties

Part1
+ A Annotations
+ ◇ Design Binder
  ³≡ Material <not specified>
+ 🎨 Lights and Cameras
+ Σ Equations
+ 🔲 Solid Bodies(1)
  ◇ Front Plane
  ◇ Top Plane
  ◇ Right Plane
  ↳ Origin
+ 🔷 Sheet-Metal1
+ 🔷 Base-Flange1
+ 🔷 Edge-Flange1
+ 🔷 Edge-Flange2
+ 🔷 Edge-Flange3
  🔷 Plane1
+ 🔲 Flat-Pattern1

In the flyout FeatureManager design tree, select **Plane1** as the **Mirror Face/Plane**. It may already be selected for you.

Under **Features to Mirror**, select **Edge-Flange1**, **Edge-Flange2**, and **Edge-Flange3** from the flyout FeatureManager design tree.

A preview will appear as you select the features to ensure that you are selecting the correct features.

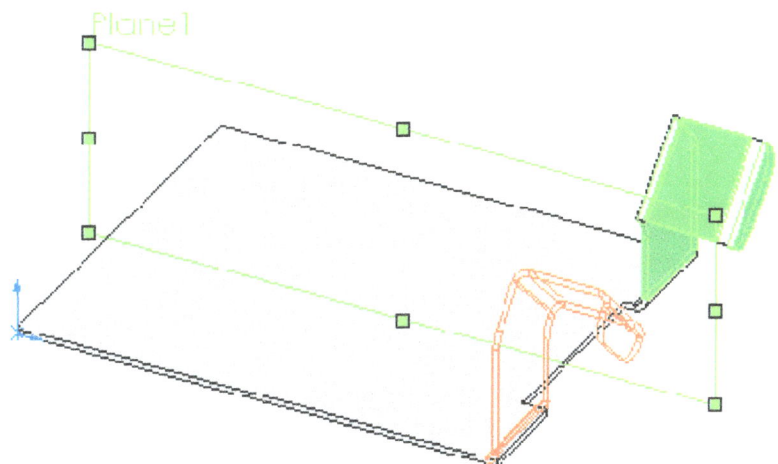

Plane1

Select the green check mark button at the top of the **Mirror** PropertyManager to accept the settings and mirror the features.

Right click on **Plane1** in the FeatureManager design tree and select **Hide** to hide the plane.

Add a Jog to the Left Side

Select the **Sheet Metal** icon in the control area of the CommandManager. Then, select the **Jog** icon in the CommandManager, or pull down the "Insert" menu and select **Sheet Metal – Jog**.

When prompted in the PropertyManager to select a planer face on which to sketch bend lines, pick the top of the base flange.

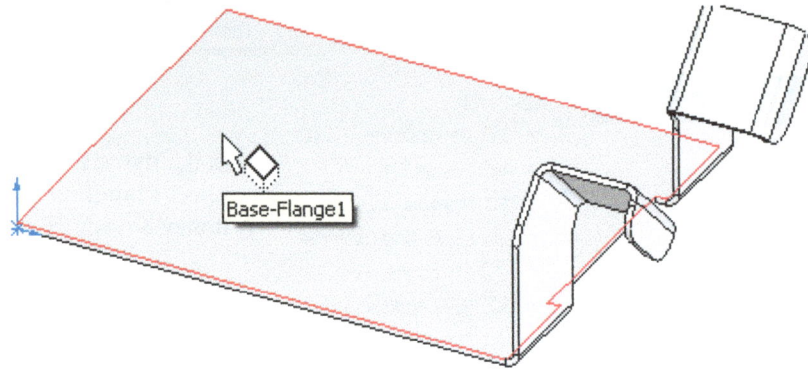

Select the **Line** icon in the CommandManager, or pull down the "Tools" menu and select **Sketch Entities – Line**.

Create a vertical line and dimension it as shown by selecting the **Smart Dimension** icon in the CommandManager, or pulling down the "Tools" menu and selecting **Dimensions – Smart**.

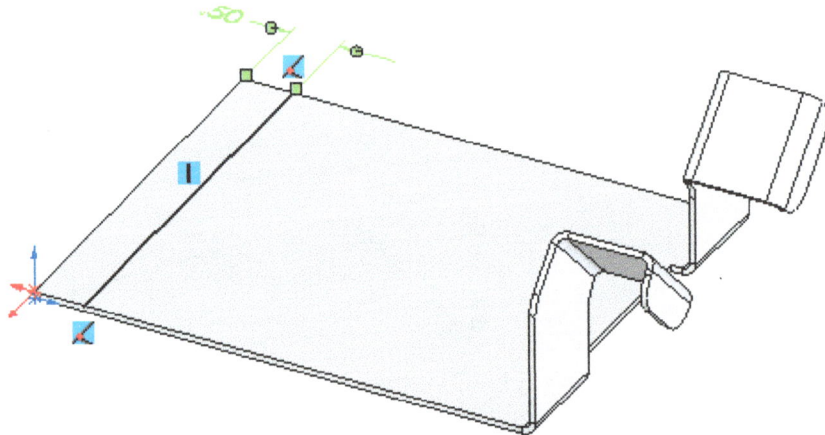

Exit the sketch by selecting the **Exit Sketch** icon in the CommandManager or in the upper right corner of the graphics area.

Select the top of Base Flange again for the **Fixed Face**. Make sure that you select to the right of the vertical line.

Set the **Jog Offset** to **Blind** distance of '**0.50in**'.

Make sure the **Dimension position Overall Dimension** button is depressed and that the **Fix projected length** check box is checked.

Also, make sure that the **Material Inside Jog Position** button is depressed and that the **Jog Angle** is set to **90.00deg**.

Select the green check mark button at the top of the **Jog** PropertyManager to accept the settings and create the jog.

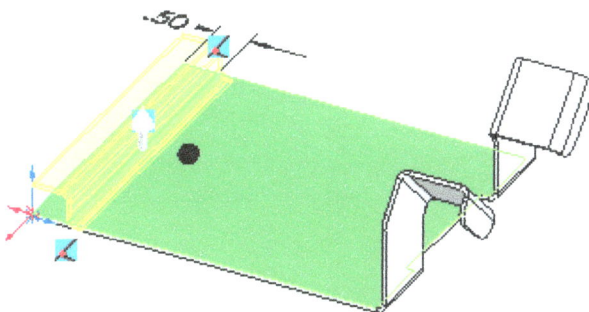

Add to the New Flange

Select the **Base-Flange/Tab** icon from the toolbar, or pull down the "Insert" menu and select **Sheet Metal – Base Flange**.

When prompted to select a plane or planar face on which to sketch the feature cross section, select the top of the jog as shown.

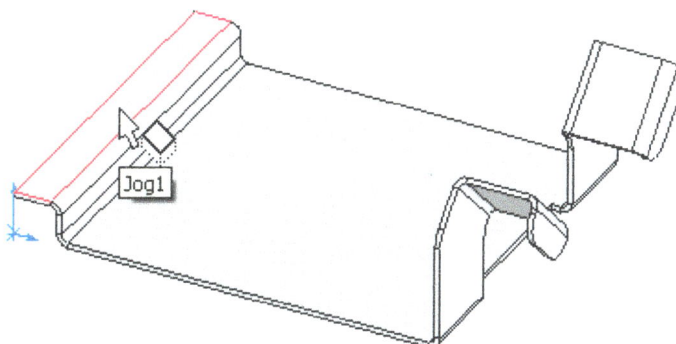

Change the display view to **Normal To** the sketch by selecting **Normal To** in the View list in the lower left corner of the graphics area.

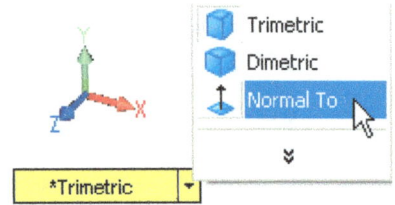

Select the **Zoom In/Out** icon in the "View" toolbar, or pull down the "View" menu and select **Modify – Zoom In/Out** to reduce the size of the part on the screen. This will give you more space to draw the lines.

Select the **Line** icon in the CommandManager, or pull down the "Tools" menu and select **Sketch Entities – Line**.

Create the lines as shown. Note that the left vertical line is created right on top of the edge line. You must create this line in order for the sketch to be a closed profile.

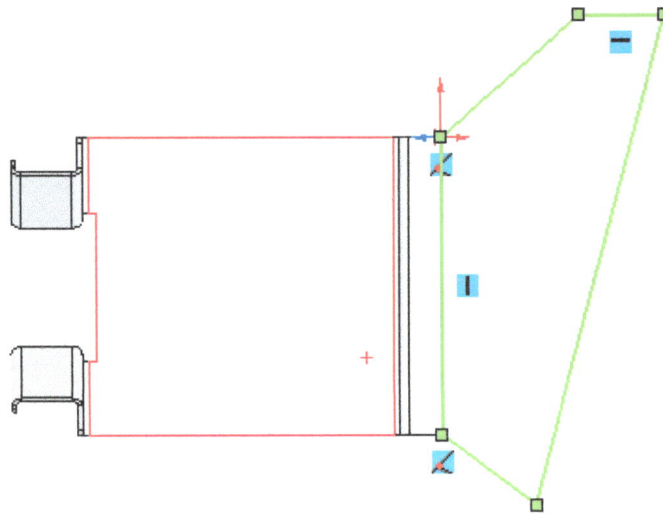

Select the **Sketch Fillet** icon in the CommandManager, or pull down the "Tools" menu and select **Sketch Tools – Fillet**.

In the **Sketch Fillet** PropertyManager, set the **Radius** to '**1.00in**'.

Select the left endpoint of the small horizontal line on the top as shown.

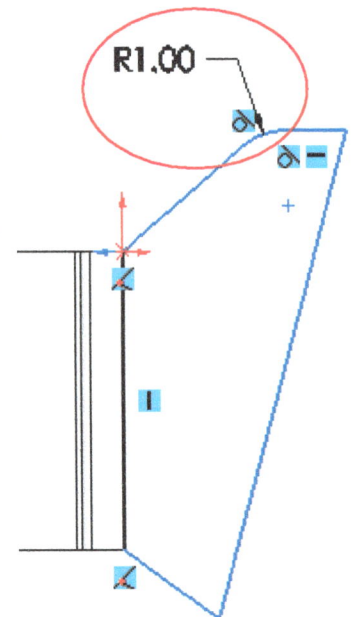

Dimension the sketch as shown by selecting the **Smart Dimension** icon in the CommandManager, or by pulling down the "Tools" menu and selecting **Dimensions – Smart**.

Exit the sketch by selecting the **Exit Sketch** icon in the CommandManager or in the upper right corner of the graphics area.

Change the display view to **Trimetric** by selecting **Trimetric** in the View list in the lower left corner of the graphics area.

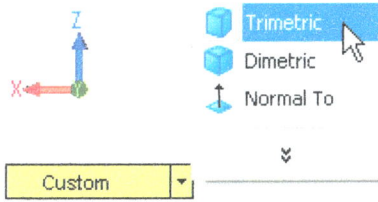

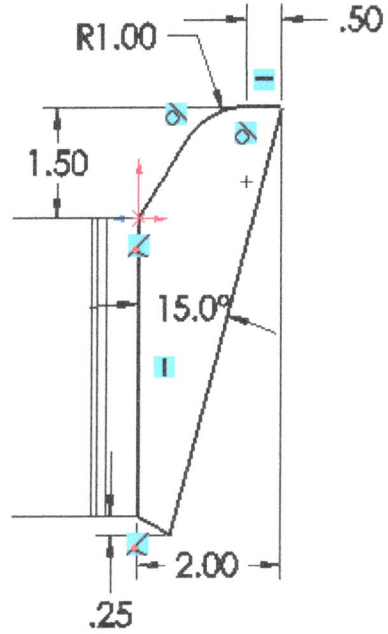

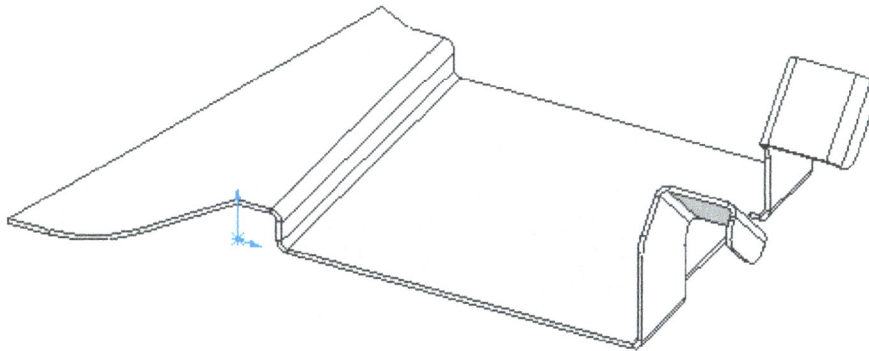

Add Fillets

Select the **Features** icon in the control area of the CommandManager. Then, select the **Fillet** icon from the toolbar, or pull down the "Insert" menu and select **Features – Fillet/Round**.

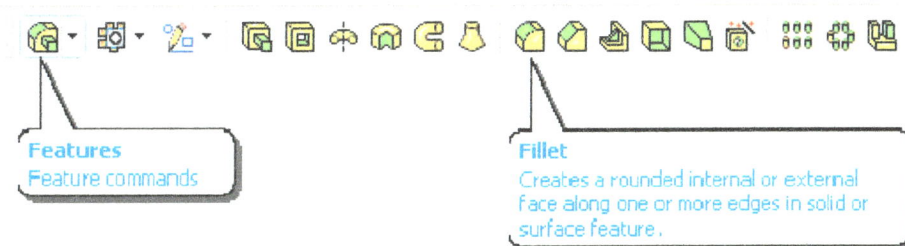

In the **Fillet** PropertyManager, set the **Radius** to '**0.25**'.

Select the two vertical edges where the end of **Jog1** and **Tab1** meet.

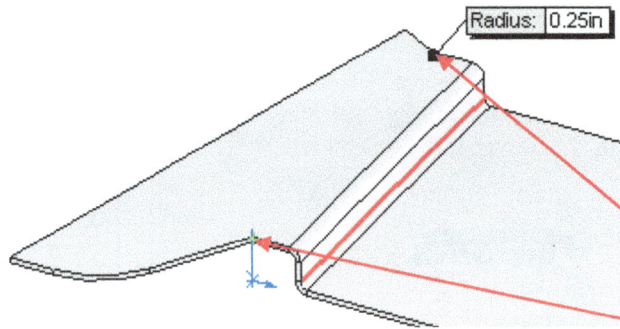

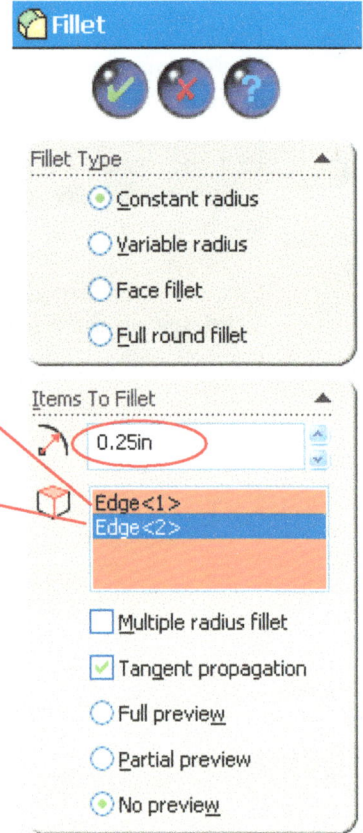

Radius: 0.25in

Fillet

Fillet Type
- ⦿ Constant radius
- ○ Variable radius
- ○ Face fillet
- ○ Full round fillet

Items To Fillet
0.25in

Edge<1>
Edge<2>

☐ Multiple radius fillet
☑ Tangent propagation
○ Full preview
○ Partial preview
⦿ No preview

You may need to use the **Zoom In/Out** icon in the "View" toolbar, or pull down the "View" menu and select **Modify – Zoom In/Out** to select the small vertical edges.

Select the green check mark button at the top of the **Fillet** PropertyManager to accept the settings and create the fillets.

One More Flange

Select the **Sheet Metal** icon in the control area of the CommandManager. Select the **Edge Flange** icon in the CommandManager, or pull down the "Insert" menu and select **Sheet Metal – Edge Flange**.

Select the left edge of the tab you just created. Move the cursor down and click to set the direction of the flange.

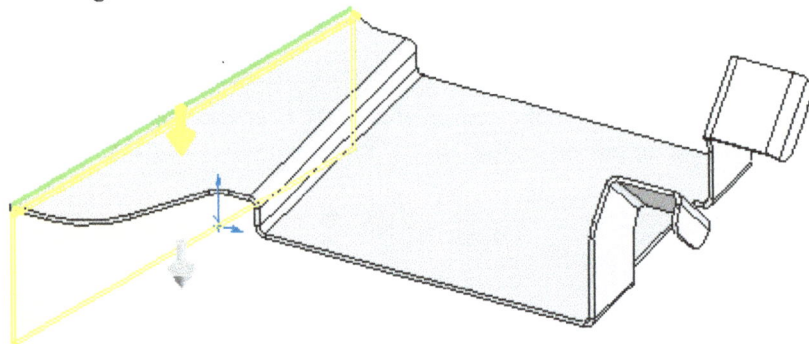

In the **Edge-Flange** PropertyManager, set the **Flange Length** to '**0.50in**', depress the **Outer Virtual Sharp** button, and depress the **Material Inside** button.

Select the green check mark button at the top of the **Edge-Flange** PropertyManager to accept the settings and create the feature.

The Last Flange

Select the **Edge Flange** icon in the CommandManager, or pull down the "Insert" menu and select **Sheet Metal – Edge Flange**.

Select the left edge of the flange just created. Move the cursor to the left and click to set the direction of the flange.

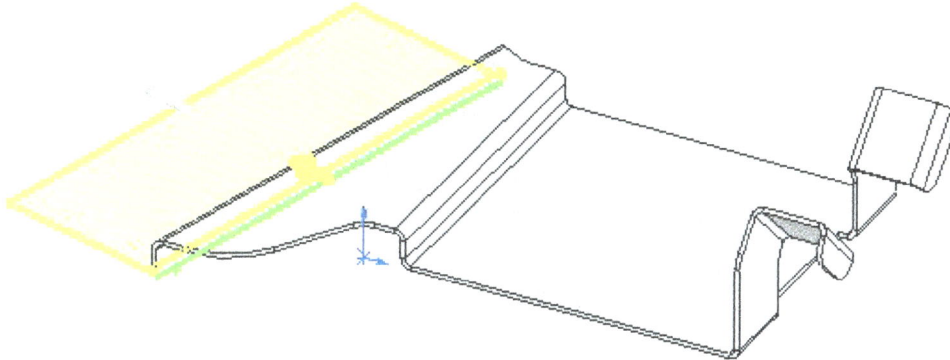

In the **Edge-Flange** PropertyManager, set the **Flange Length** to '1.00in', depress **the Outer Virtual Sharp** button, and depress the **Material Inside** button.

Select the green check mark button at the top of the **Edge-Flange** PropertyManager to accept the settings and create the last flange.

Add Corner Breaks

Select the **Break-Corner/Corner-Trim** icon in the CommandManager, or pull down the "Insert" menu and select **Sheet Metal – Break Corner**.

In the **Break Corner** PropertyManager, select the **Fillet Break type** button and set the **Radius** to '0.50'.

Select the top face of the last flange as shown.

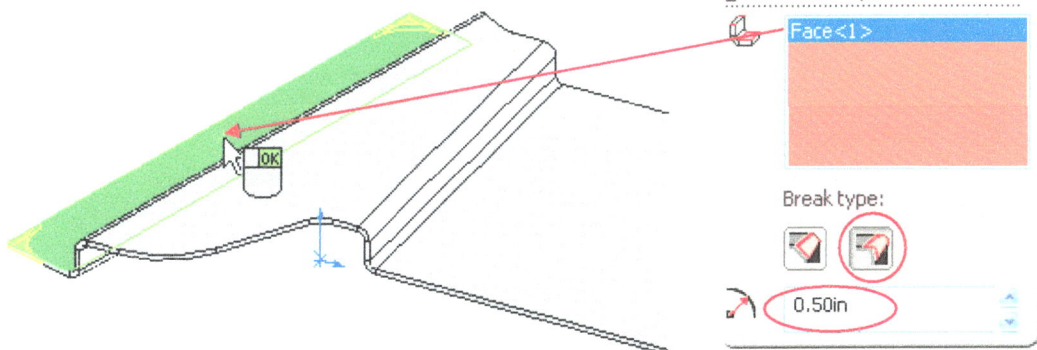

Select the green check mark button at the top of the **Break Corner** PropertyManager to accept the settings and create the corner breaks.

Saving the Part

Select the **Save** icon in the "Standard" toolbar, or select **Save** from the "File" pull down menu.

The **Save As** dialog box appears.

In the **File name** box, type the name of the drawing number. For this lesson, use '**4589-854**' or any other name you wish to assign it, and select **Save**.

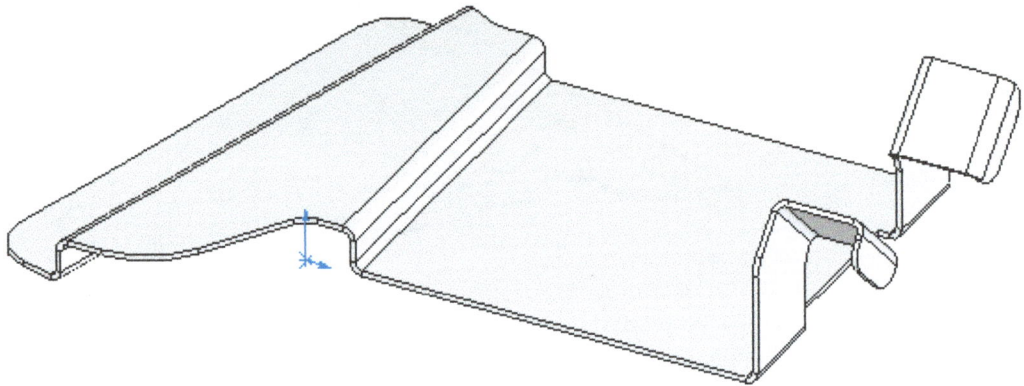

Checking for Part Overlap in the Flat Pattern

There is a potential situation that you need to be aware of. The part may look correct while folded up. However, the flanges may actually overlap once the part is flattened.

To look for this, select the **Flatten** icon in the Sheet Metal CommandManager. Or you may also right click on **Flat-Pattern1** in the FeatureManager design tree and select **Unsuppress**.

Flatten
Shows the flat pattern for the existing sheet metal part.

Change the display view to a top view by selecting **Top** in the View list in the lower left corner of the graphics area.

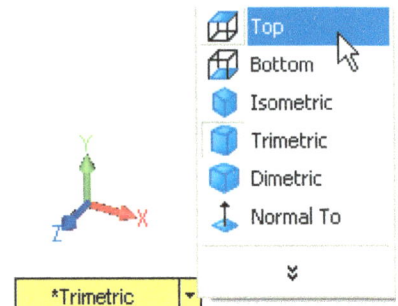

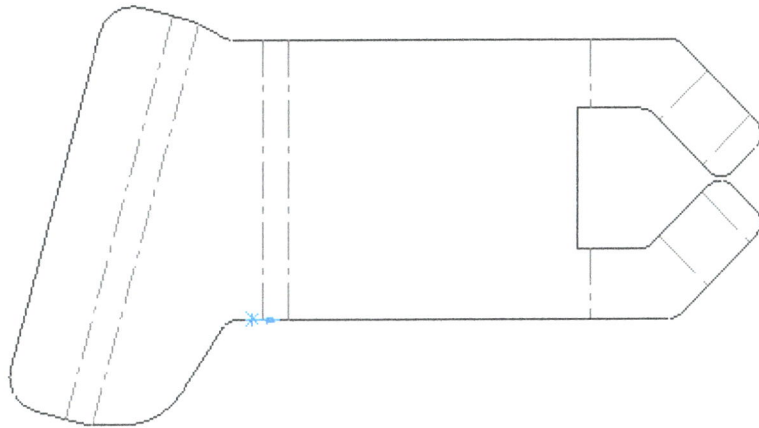

You will notice that the tabs on the right side of the part do not overlap. Just enough room to get the laser through them. But what if the base flange was a smaller size?

Right click on **Base-Flange1** in the FeatureManager design tree and select **Edit Sketch**.

Change the height dimension from '**4.00**' to '**3.00**' by double clicking on the dimension and changing the value of the dimension.

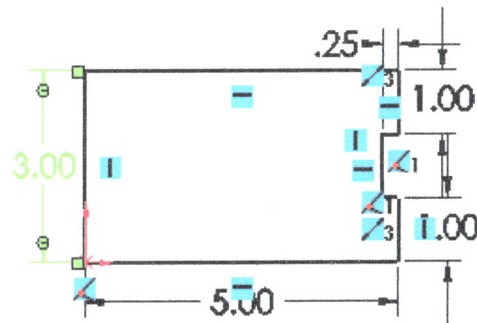

Select the **Exit Sketch** icon in the CommandManager or in the upper right corner of the graphics area.

Now, the flanges do overlap.

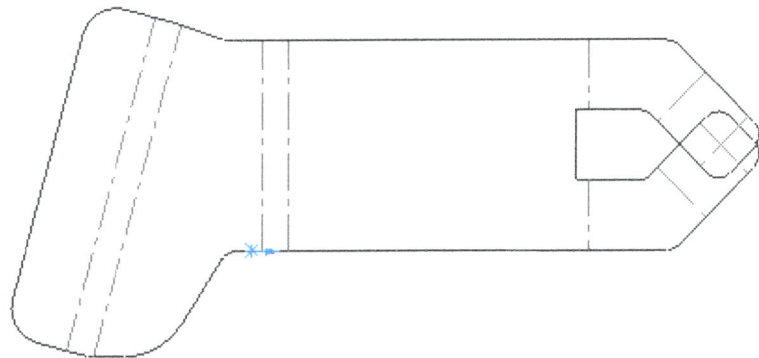

To refold the part, select the **Flatten** icon in the CommandManager toolbar again to fold the part back up. You may also right click on **Flat-Pattern1** in the FeatureManager design tree and select **Suppress**.

Select the **Previous View** icon in the "View" toolbar, or pull down the "View" menu and select **Modify – Previous View** to return to the Trimetric view.

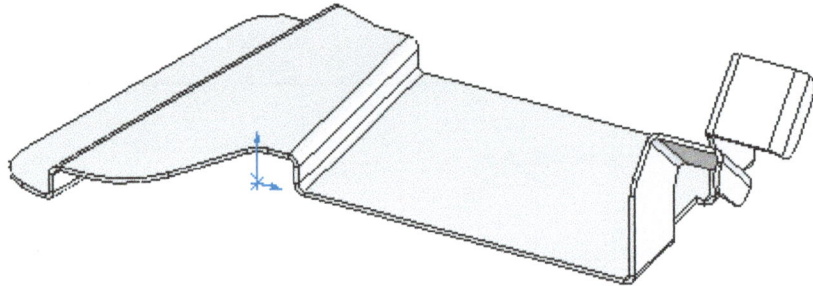

There are no error messages indicating that there is a problem with the part. The only way to check for the part overlap is to manually look at it in the flat pattern.

Closing the File

Select **Close** from the "File" pull down menu.

Select **No** when prompted to **Save changes to 4589-854.SLDPRT?** to not save your changes.

Chapter 8

Panel Cover

The Panel Cover is really a simple part, except that the hold dimension is inside of the internal flanges. This is easily accomplished by adding an equation after the part is created.

The Unfold and Fold command are introduced to make certain that the space is correct around the internal flanges for a .25 wide punch. This is a very handy technique and will be further discussed in Course 3.

Another new feature used with this part is the Link Value command. This allows you to set several dimensions all equal to the same value. A great tool to make your future editing simpler.

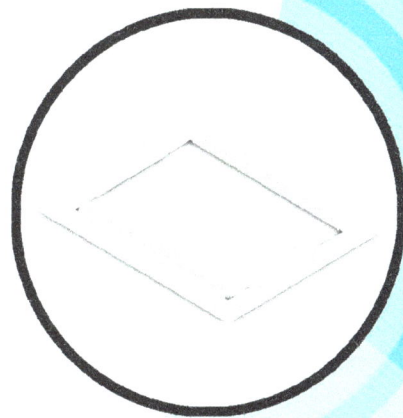

Create the Base Flange

Start by creating a new part document by selecting the **New** icon in the "Standard" toolbar, or pull down the "File" menu and select **New**.

Select **Part** in the **New SolidWorks Document** dialog box and then select **OK**.

The first feature that you will create in this part is a base flange created from a sketched rectangular profile.

To do this, select the **Sheet Metal** icon in the control area of the CommandManager. Then, select the **Base-Flange/Tab** icon from the toolbar, or pull down the "Insert" menu and select **Sheet Metal – Base Flange**.

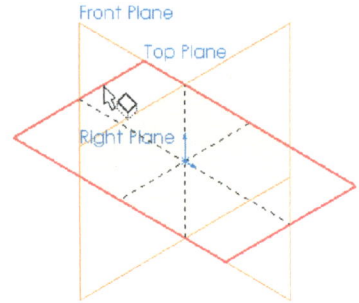

Select the **Top** plane when prompted to select a plane on which to sketch the feature cross-section.

Create a rectangle as shown below with the lower left hand corner starting at the origin using the **Rectangle** icon in the CommandManager, or pull down the "Tools" menu and select **Sketch Entities – Rectangle**.

Create second rectangle inside of the first rectangle as shown below.

Dimension the sketch as shown by selecting the **Smart Dimension** icon in the CommandManager, or pull down the "Tools" menu and select **Dimensions – Smart**. If the smaller rectangle moves to be partially outside of the larger rectangle, just pick the line which is outside and drag it back to the inside.

You may need to use the **Zoom to Fit** icon in the "View" toolbar, or pull down the "View" menu and select **Modify – Zoom to Fit** so you can see the entire rectangle.

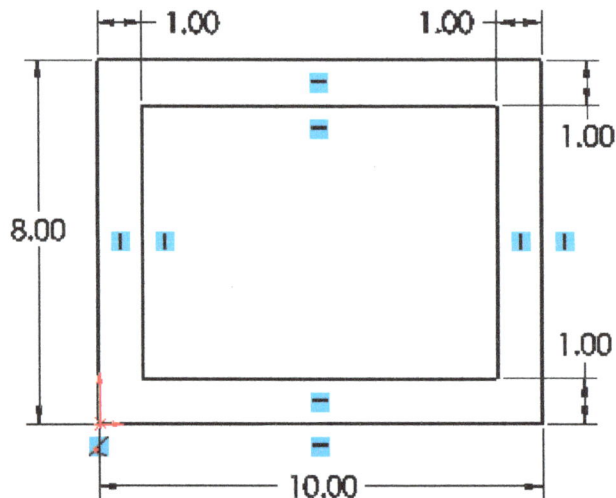

Exit the sketch by selecting the **Exit Sketch** icon in the CommandManager or in the upper right corner of the graphics area.

Complete the Base Flange

In the **Base Flange** PropertyManager, under **Sheet Metal Parameters**, set the **Thickness** to '**0.036in**', and check the **Reverse direction** check box.

Select the green check mark button at the top of the **Base Flange** PropertyManager to accept the settings and create the base flange.

Add a Hem

Select the **Hem** icon in the CommandManager, or pull down the "Insert" menu and select **Sheet Metal – Hem**.

Select the four outside edges of the part as shown.

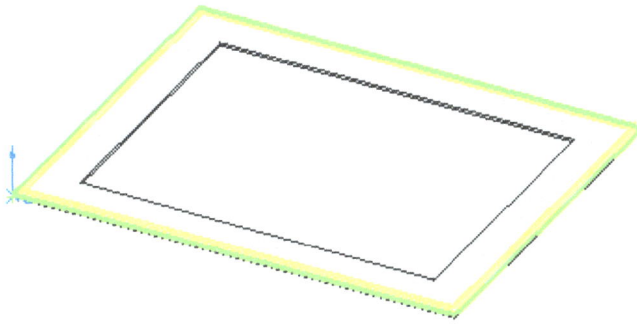

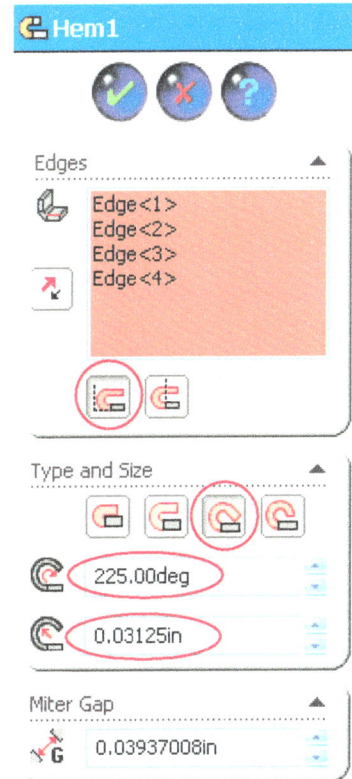

Select the **Material Inside** button.

Under **Type and Size**, select the **Tear Drop** button.

Set the **Angle** to '**225.00deg**' and set the **Radius** to '**0.03125in**'.

Select the green check mark button at the top of the **Hem** PropertyManager to accept the settings and create the feature.

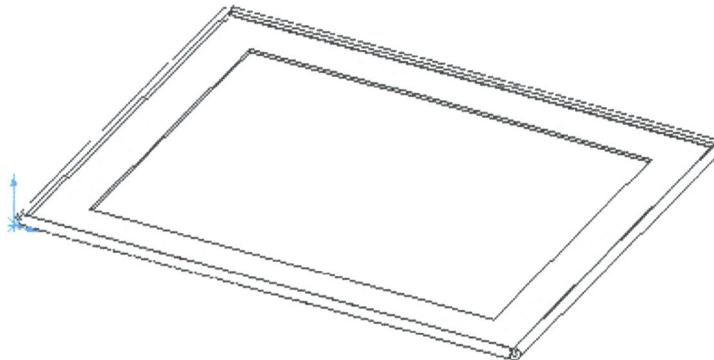

Create an Edge Flange

Select the **Edge Flange** icon in the CommandManager, or pull down the "Insert" menu and select **Sheet Metal – Edge Flange**.

Select the top inside edge as the edge to add the flange to as shown, and move the cursor up and click to set the direction of the flange.

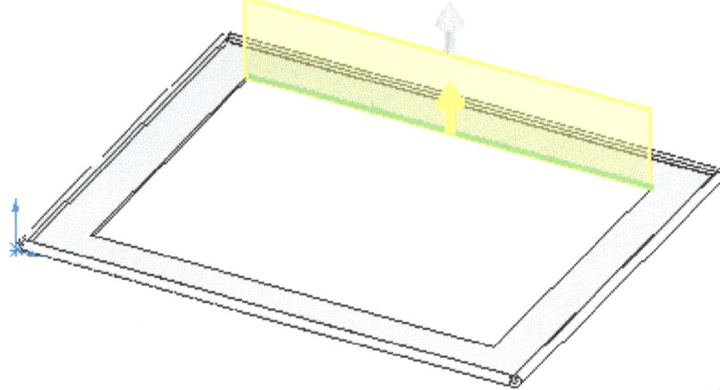

In the **Edge-Flange** PropertyManager, select the **Edit Flange Profile** button.

Left click and drag the sides of the rectangle in toward the middle.

Select the **Smart Dimension** icon and add a '**0.25**' dimension from the vertical line to the small vertical edge line on each side as shown.

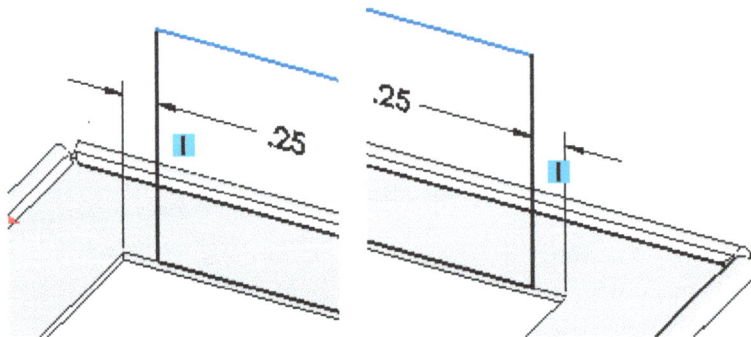

Select the **Back** button in the **Profile Sketch** dialog box.

In the **Edge-Flange** PropertyManager, set the **Flange Length** to a **Blind** distance of '**0.75in**'. Make sure that the **Outer Virtual Sharp** button is depressed. Set the **Flange Position** to **Bend Outside**.

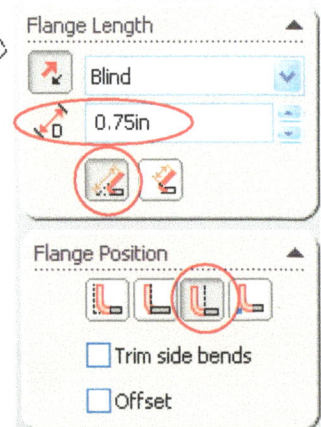

Select the green check mark button at the top of the **Edge-Flange** PropertyManager to accept the settings and create the feature.

Add Another Edge Flange

Select the **Edge Flange** icon in the CommandManager, or pull down the "Insert" menu and select **Sheet Metal – Edge Flange**.

Select the top edge of the flange you just created and move the cursor up and to the right and click to set the direction of the flange.

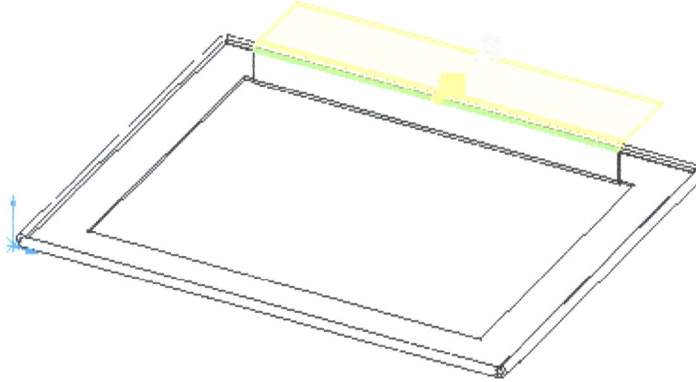

In the **Edge-Flange** PropertyManager, set the **Flange Length** to a **Blind** distance of '**0.375in**'. Make sure that the **Outer Virtual Sharp** button is depressed. Set the **Flange Position** to **Material Inside**.

Select the green check mark button at the top of the **Edge-Flange** PropertyManager to accept the settings and create the flange.

In the FeatureManager design tree, right click on the **Sheet-Metal1** item and select **Edit Feature** from the menu.

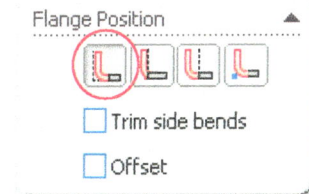

In the **Sheet-Metal1** PropertyManager, enter '**0.075in**' for the default **Bend Radius**.

Select the green check mark button at the top of the **Sheet-Metal1** PropertyManager.

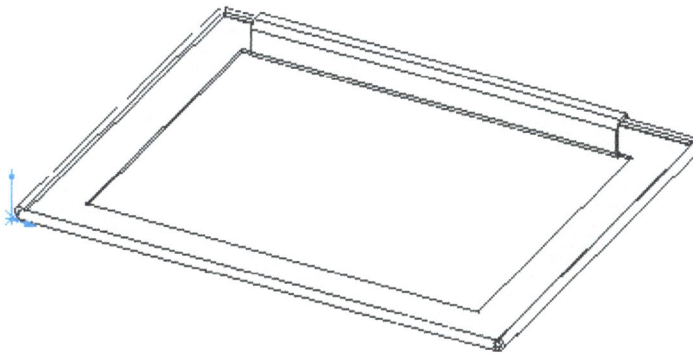

Unfold a Single Flange

In order to control the width of the adjacent edge flange, we must first unbend the previous edge flanges.

Select the **Unfold** icon in the CommandManager, or pull down the "Insert" menu and select **Sheet Metal – Unfold**.

Unfold
Unfolds bends in a sheet metal part.

For the **Fixed face**, select the top face of the part.

For the **Bends to unfold**, select the two bend areas from the edge flanges.

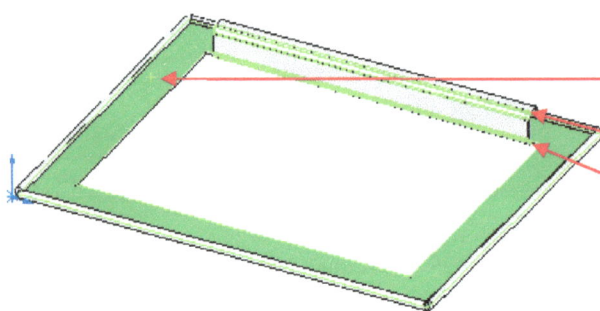

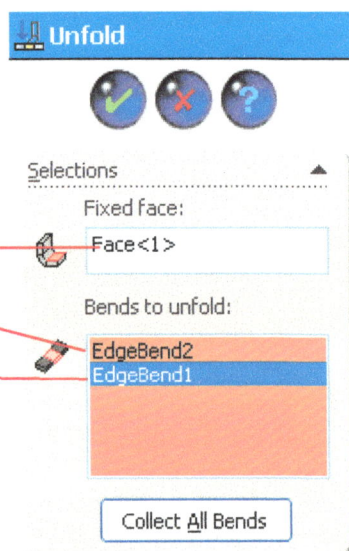

Select the green check mark button at the top of the **Unfold** PropertyManager and the bends will unfold.

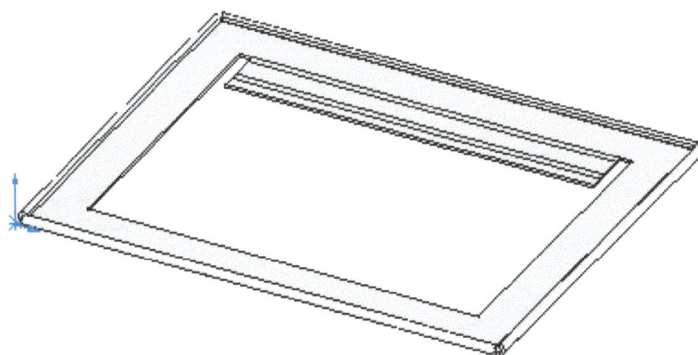

Add an Edge Flange on the Left

Select the **Edge Flange** icon in the CommandManager, or pull down the "Insert" menu and select **Sheet Metal – Edge Flange**.

Select the left inner edge of the base flange and move the cursor up and click to set the direction of the flange.

In the **Edge-Flange** PropertyManager, select the **Edit Flange Profile** button.   Edit Flange Profile

Left click and drag both sides of the rectangle in toward the middle as shown below. Move the right side of the rectangle past the center of the part.

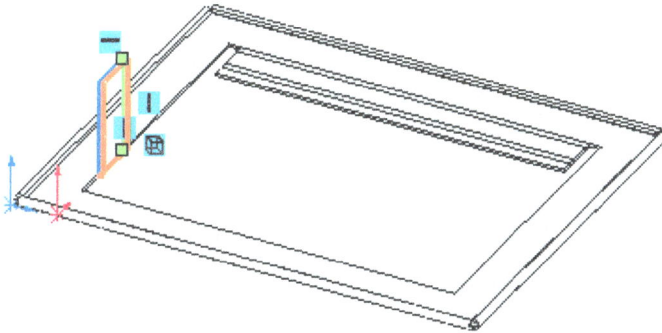

Select the **Centerline** icon in the CommandManager, or pull down the "Tools" menu and select **Sketch Entities – Centerline**.

Create a vertical centerline at the midpoint of the edge line as shown. Make sure that you get the midpoint of the edge line. The edge line will highlight when its midpoint is shown.

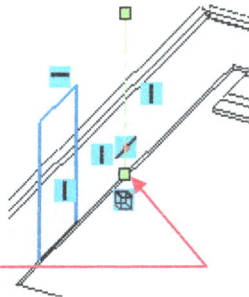

Next, select the **Smart Dimension** icon and add a '**0.25**' dimension from the unfolded flange to the right side of the rectangle sketch as shown. Select the thickness edge of the unfolded flange and the right side of the rectangle sketch to add the dimension.

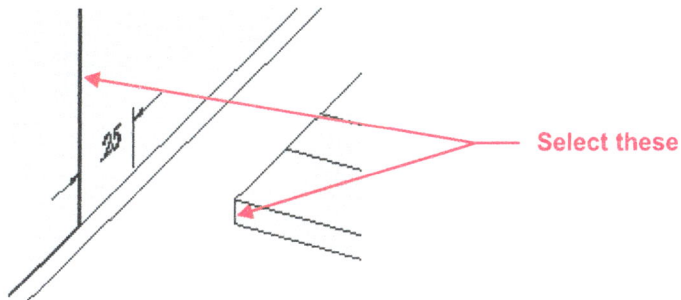

Select these

Select the **Add Relation** icon in the CommandManager, or pull down the "Tools" menu and select **Relations – Add**.

Select the centerline and the two vertical sketch lines.

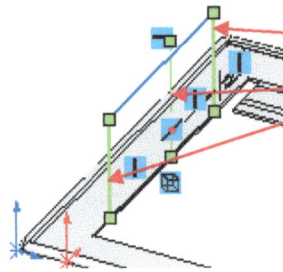

Selected Entities

Line3
Line5
Line7

Existing Relations

Add Relations

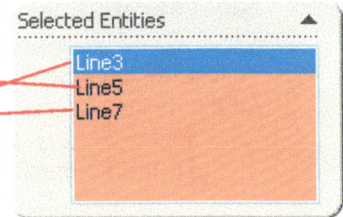

In the **Add Relations** PropertyManager, select the **Symmetric** button.

Select the green check mark button at the top of the **Add Relations** PropertyManager.

Select the **Back** button in the **Profile Sketch** dialog box.

< Back

— Horizontal

| Vertical

Collinear

Parallel

= **Equal**

Symmetric

Fix

In the **Edge-Flange** PropertyManager, set the **Flange Length** to a **Blind** distance of '**0.75in**'. Make sure that the **Outer Virtual Sharp** button is depressed. Set the **Flange Position** to **Bend Outside**.

Select the green check mark button at the top of the **Edge-Flange** PropertyManager to accept the settings and create the feature.

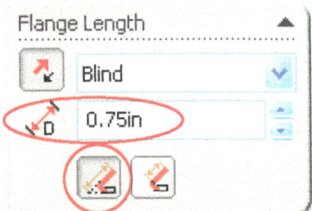

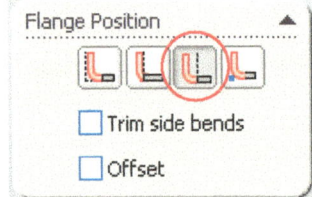

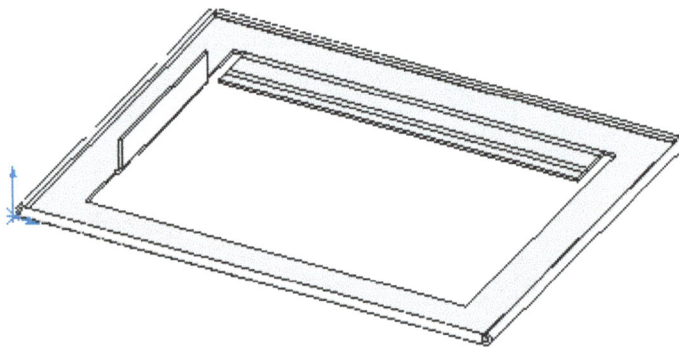

Flange Length

Blind

0.75in

Flange Position

Trim side bends

Offset

Add Another Edge Flange on the Left

Select the **Edge Flange** icon in the CommandManager, or pull down the "Insert" menu and select **Sheet Metal – Edge Flange**.

Select the top edge of the flange just created and move the cursor up and to the left and click to set the direction of the flange.

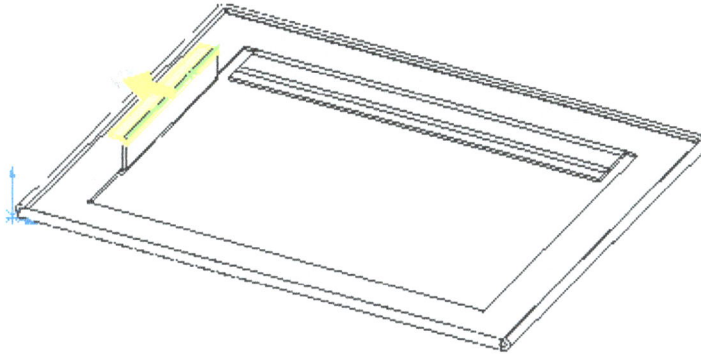

In the **Edge-Flange** PropertyManager, set the **Flange Length** to a **Blind** distance of '**0.375in**'. Make sure that the **Outer Virtual Sharp** button is depressed. Set the **Flange Position** to **Material Inside**.

Select the green check mark button at the top of the **Edge-Flange** PropertyManager to accept the settings and create the flange.

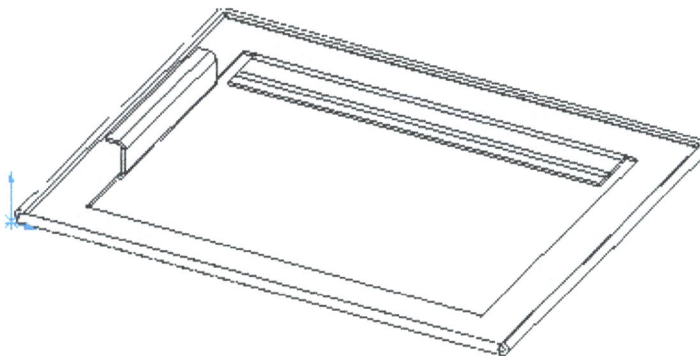

Folding the Flange Back Up

Select the **Fold** icon in the CommandManager, or pull down the "Insert" menu and select **Sheet Metal – Fold**.

Fold
Folds flattened bends in a sheet metal part.

For the **Fixed face**, select the top face of the part.

In the **Fold** PropertyManager, select the **Collect All Bends** button.

Select the green check mark button at the top of the **Fold** PropertyManager and the bends will be folded.

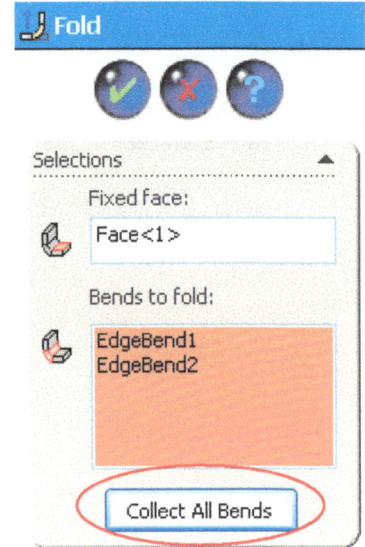

Mirroring the Top Flange

Pull down the "Insert" menu, and select **Reference Geometry – Plane**.

In the upper left hand corner of the graphics screen, click on the plus sign next to the document name (Part1) to expand the flyout FeatureManager design tree.

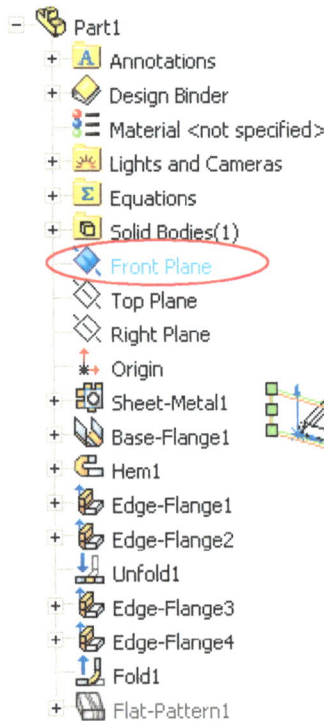

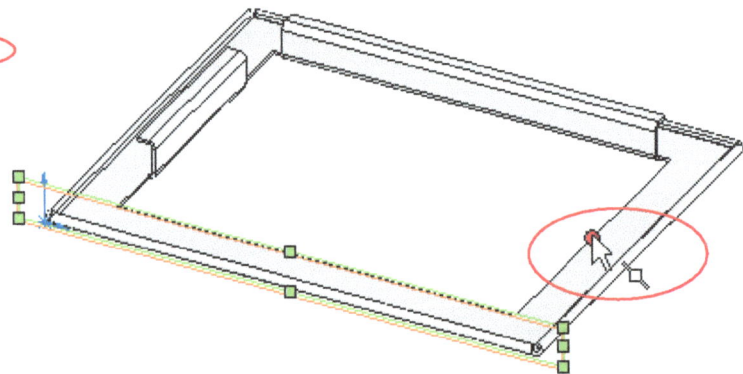

Left click on **Front Plane**. Then, in the graphics area, select the midpoint of the inner right-hand edge of the base flange as shown.

In the **Plane** PropertyManager, select the **Parallel Plane at Point** button. This may already be selected for you.

A parallel plane is created for you at the selected point.

Select the green check mark button at the top of the **Plane** PropertyManager.

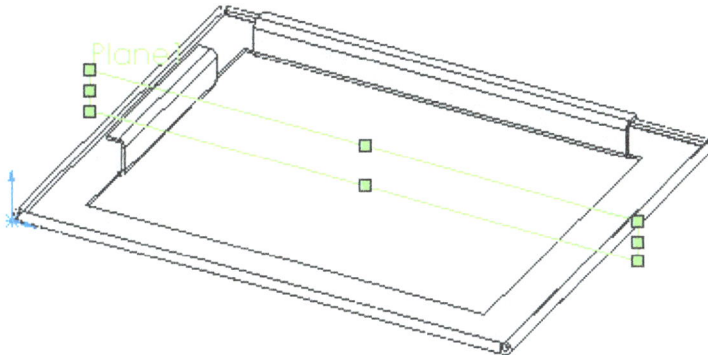

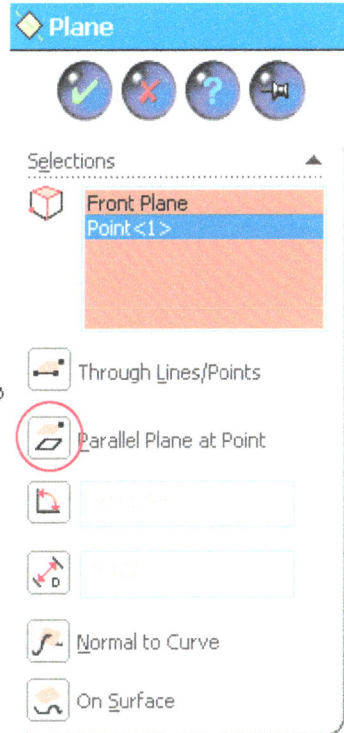

Select the **Features** icon in the control area of the CommandManager. Then, select the **Mirror** icon from the toolbar, or pull down the "Insert" menu and select **Pattern/Mirror – Mirror**.

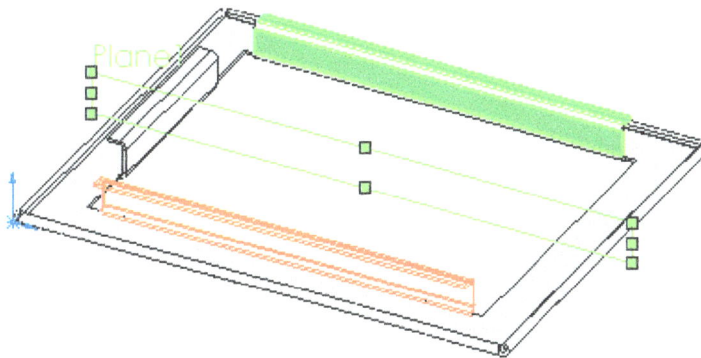

Select **Plane1** as the **Mirror Face/Plane**.

Under **Features to Mirror**, select **Edge-Flange1** and **Edge-Flange2** from the flyout FeatureManager design tree.

A preview will appear as you select the features to ensure that you are selecting the correct features.

Select the green check mark button at the top of the **Mirror** PropertyManager to accept the settings and create the feature.

In the FeatureManager design tree, right click on **Plane1** and select **Hide** to hide the plane.

Repeat the Mirror for the Left Flange

Pull down the "Insert" menu, and select **Reference Geometry – Plane**.

In the flyout FeatureManager design tree, left click on **Right Plane**. Then, in the graphics area, select the midpoint of the front inner edge of the base flange as shown.

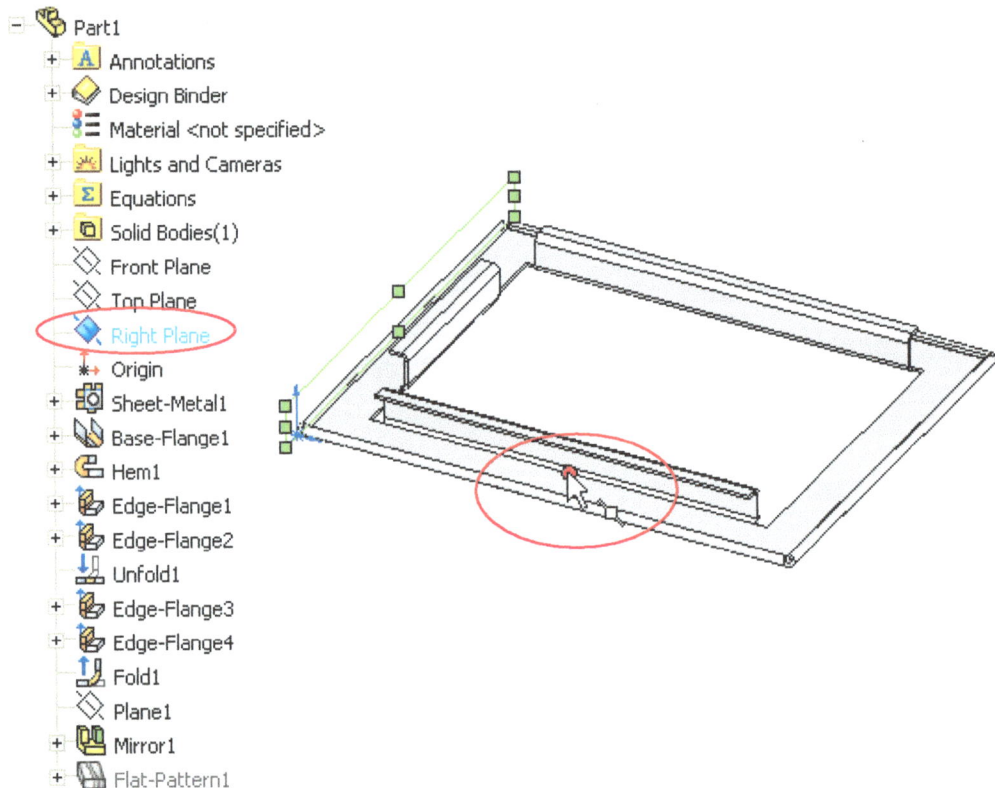

In the **Plane** PropertyManager, select the **Parallel Plane at Point** button. Again this may already be selected for you.

A parallel plane is created for you at the selected point.

Select the green check mark button at the top of the **Plane** PropertyManager.

Select the **Mirror** icon from the toolbar, or pull down the "Insert" menu and select **Pattern/Mirror – Mirror**.

Mirror
Mirrors features, faces, and bodies about a face or a plane.

Select **Plane2** as the **Mirror Face/Plane**.

Under **Features to Mirror**, select **Edge-Flange3** and **Edge-Flange4** from the flyout FeatureManager design tree.

- Part1
  - + Annotations
  - + Design Binder
  - Material <not specified>
  - + Lights and Cameras
  - + Equations
  - + Solid Bodies(1)
  - Front Plane
  - Top Plane
  - Right Plane
  - Origin
  - + Sheet-Metal1
  - + Base-Flange1
  - + Hem1
  - + Edge-Flange1
  - + Edge-Flange2
  - Unfold1
  - + Edge-Flange3
  - + Edge-Flange4
  - Fold1
  - Plane1
  - + Mirror1
  - Plane2
  - + Flat-Pattern1

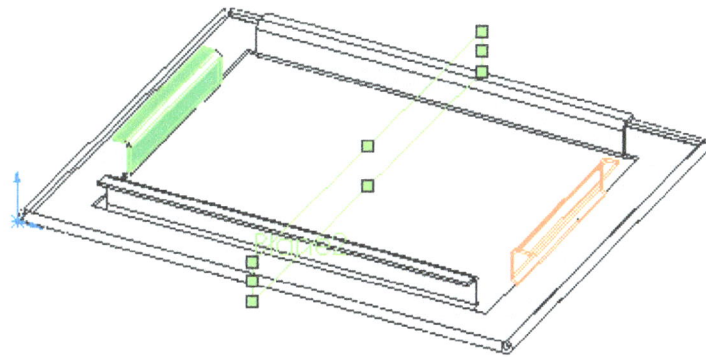

Select the green check mark button at the top of the **Mirror** PropertyManager to accept the settings and create the feature.

In the FeatureManager design tree, right click on **Plane2** and select **Hide** to hide the plane.

Applying the Design Intent

The design intent is that the inside to inside of the flanges should be 6 inches and 8 inches respectively. The Base Flange sketch was used to establish this by adding 1 inch dimensions between the outside and inside rectangle. But then we created the interior flanges, causing the part dimensions to be off by a bend radius and a thickness.

To adjust the part so that it meets the design intent, right click in the FeatureManager design tree on the **Equations** folder and select **Add Equation**.

In the **Add Equation** dialog box, type '**SideWidth** = 1– (**Thickness +**'.

Then, in the FeatureManager design tree, double click on **Sheet-Metal1** so that the radius dimension is shown. Click on the radius dimension ("D1@Sheet-Metal1").

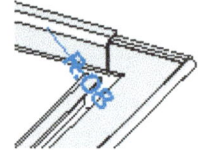

Finally, in the **Add Equation** dialog box, press the '**)**' button. The exact equation is shown in the dialog box below.

Select the **OK** button to add the equation.

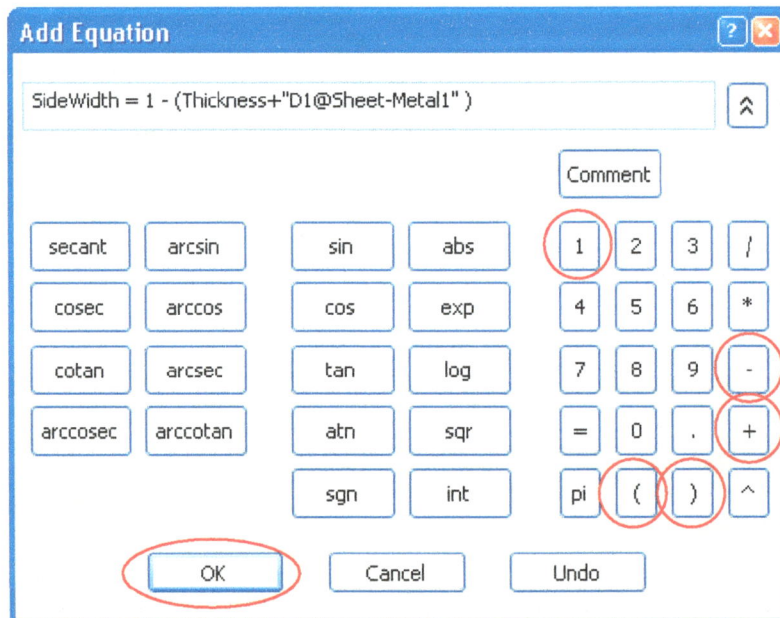

In the **Equations** dialog box, select the **OK** button.

In the FeatureManager design tree, right click on **Base-Flange1** and select **Edit Sketch**.

Double click on the upper **1.00** dimension. In the **Modify** dialog box, select **Link Value** as shown.

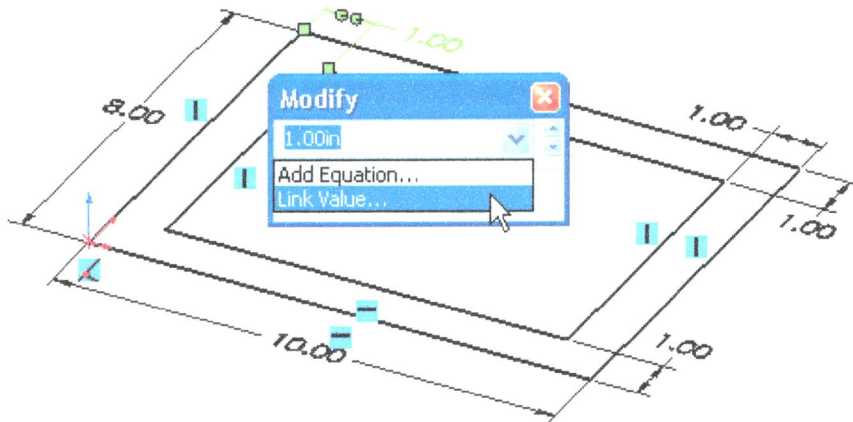

In the Shared Values dialog box, pull down the "Name" menu and select '**$VAR: SideWidth**'. Then, select the **OK** button.

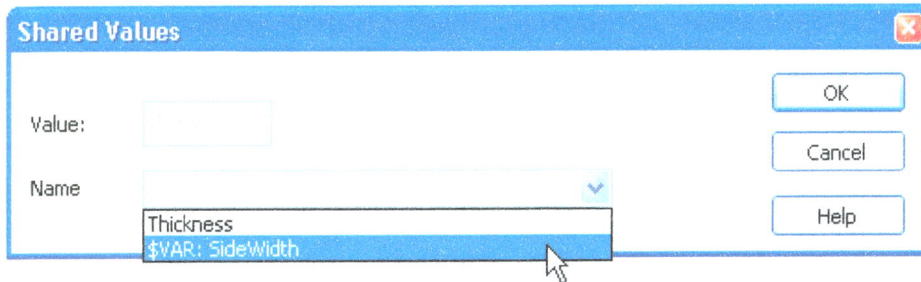

Repeat the above command to link the other three **1.00** dimensions. Now, anytime the thickness or bend radius changes, the desired dimensions are held.

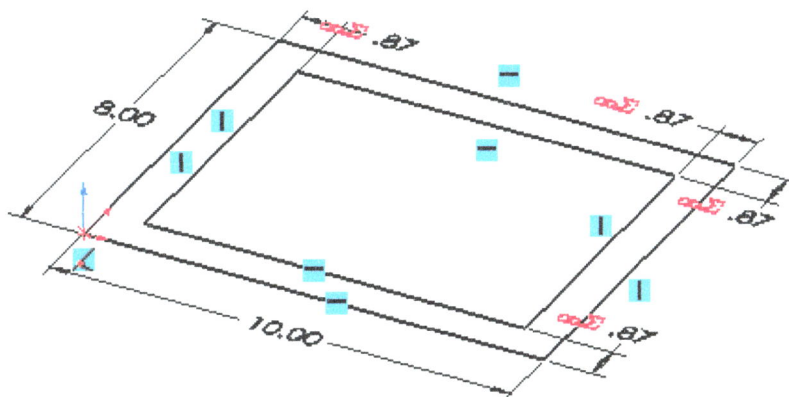

Exit the sketch by selecting the **Exit Sketch** icon in the CommandManager or in the upper right corner of the graphics area.

Checking for Part Overlap in the Flat Pattern

Select the **Flatten** icon from the Sheet Metal CommandManager toolbar. You may also right click on **Flat-Pattern1** and select **Unsuppress**.

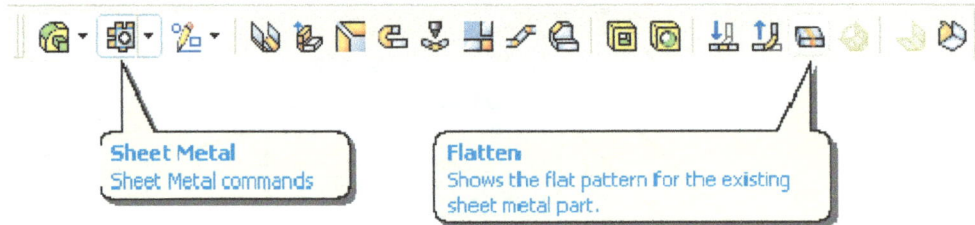

Change the display view to **Top** by selecting **Top** in the View list in the lower left corner of the graphics area. You will notice that the gap is still **0.25** between the tabs.

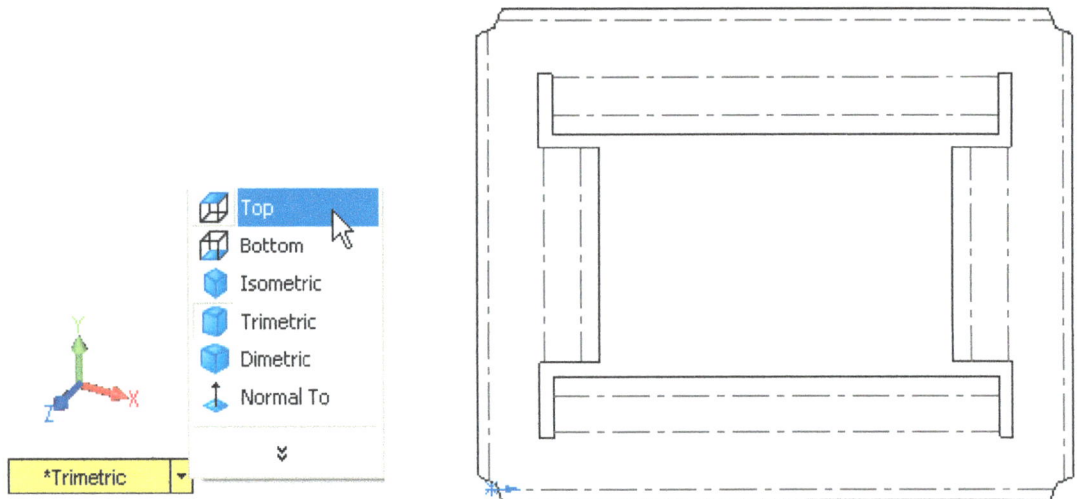

Saving the Part

Select the **Save** icon in the "Standard" toolbar, or select **Save** from the "File" pull down menu.

The **Save As** dialog box appears.

In the **File name** box, type the name of the drawing number. For this lesson, use '**PanelCover**' and select **Save**.

Closing the File

Select **Close** from the "File" pull down menu. Select **No** when prompted to **Save changes to PanelCover.SLDPRT?**

Chapter 9

Family of Parts

The purpose of this tutorial is to create a Z-Bracket based on three key dimensions. Changes to the material thickness or the bend radius will not affect these key dimensions. Also, these three dimensions may be altered to create a family of parts from the Z-Bracket.

Here are the requirements of the Z-Bracket:

- The height dimension must be held to a specific value.
- The width of the top flange dimension must be held to a specific value.
- The outside to outside width of the part must be held to a specific value.
- The length of the bracket can be adjusted as desired.

Changes to the material thickness or the bend radius will not affect the three key dimensions.

Configurations provide an easy way to create and manage a family of parts, or assemblies, within a single file. Multiple variations of the Z-Bracket will be created by changing key dimensions.

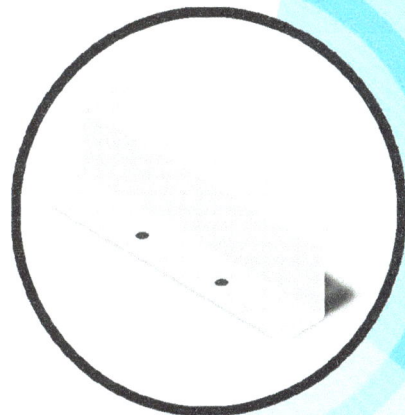

Create the Base Flange

Begin by creating a new part document by selecting the **New** icon in the "Standard" toolbar, or pull down the "File" menu and select **New**.

Select **Part** in the **New SolidWorks Document** dialog box and then select **OK**.

Create a base flange from a sketched rectangular profile by selecting the **Sheet Metal** icon in the control area of the CommandManager. Then, select the **Base-Flange/Tab** icon from the toolbar, or pull down the "Insert" menu and select **Sheet Metal – Base Flange**.

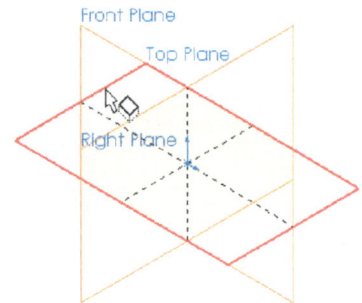

Select the **Top** plane when prompted to select a plane on which to sketch the feature cross-section.

Create a rectangle with the lower left hand corner starting at the origin using the **Rectangle** icon in the CommandManager, or pull down the "Tools" menu and select **Sketch Entities – Rectangle**.

Add a '**12.00**' horizontal dimension to the top line and a '**4.00**' vertical dimension to the right vertical line using the **Smart Dimension** icon in the CommandManager, or pull down the "Tools" menu and select **Dimensions – Smart**.

Exit the sketch by selecting the **Exit Sketch** icon in the CommandManager or in the upper right corner of the graphics area.

In the **Base Flange** PropertyManager under **Sheet Metal Parameters**, set the **Thickness** to '**0.06in**', and check the **Reverse direction** check box.

Select the green check mark button at the top of the **Base Flange** PropertyManager to accept the settings and create the base flange.

Creating the Bends

Select the **Jog** icon in the CommandManager, or pull down the "Insert" menu and select **Sheet Metal – Jog**.

Select the top of the part.

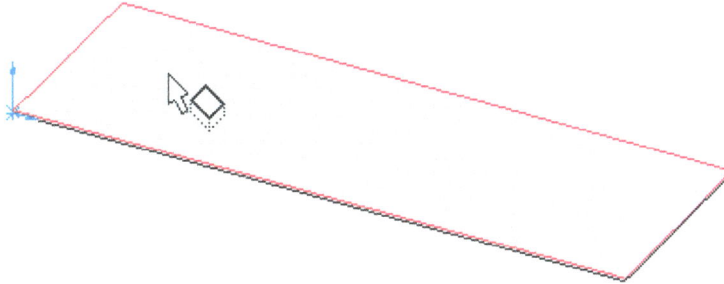

Select the **Line** icon in the CommandManager, or pull down the "Tools" menu and select **Sketch Entities – Line**.

Create a horizontal line on the top of the part. To show that the line is horizontal, the cursor changes to

Select the **Smart Dimension** icon in the CommandManager, or pull down the "Tools" menu and select **Dimensions – Smart**.

Select the left endpoint of the line that you created and the upper left corner of the part to add a '**2.00**' vertical dimension.

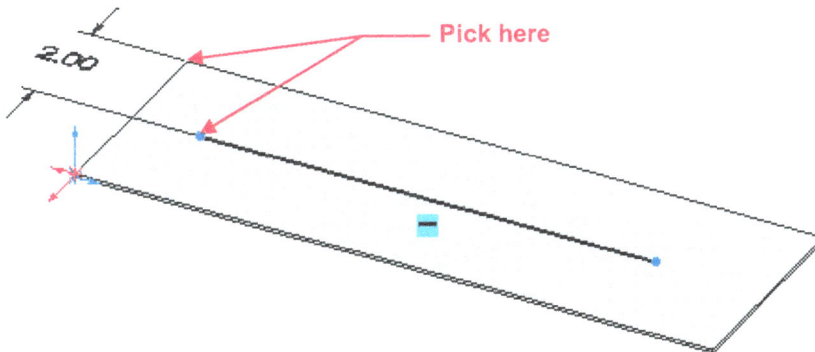

Exit the sketch by selecting the **Exit Sketch** icon in the CommandManager or in the upper right corner of the graphics area.

In the graphics area, pick the top face of the part below the horizontal line. In the preview of the Jog feature, a black bullet appears where you selected, indicating the fixed face.

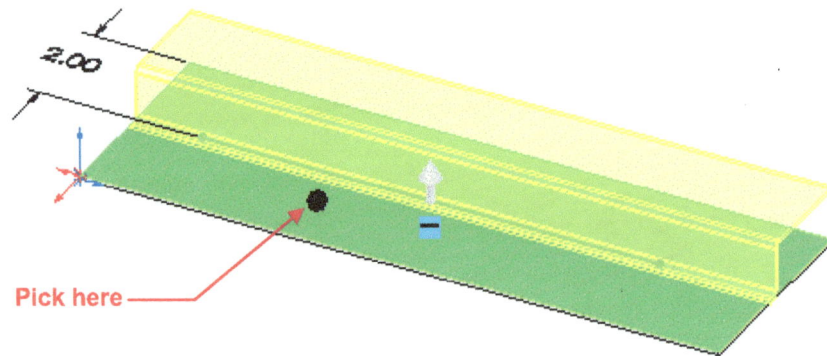

Pick here

In the **Jog** PropertyManager, under **Jog Offset**, make sure that the **End Condition** is set to **Blind** and enter '**3.00**' for the **Offset Distance**.

Make sure that the **Overall Dimension** button depressed, the **Material Outside** button is depressed, and the **Jog Angle** is set to '**90.00deg**'.

Select the green check mark button at the top of the **Jog** PropertyManager to accept the settings and create the feature.

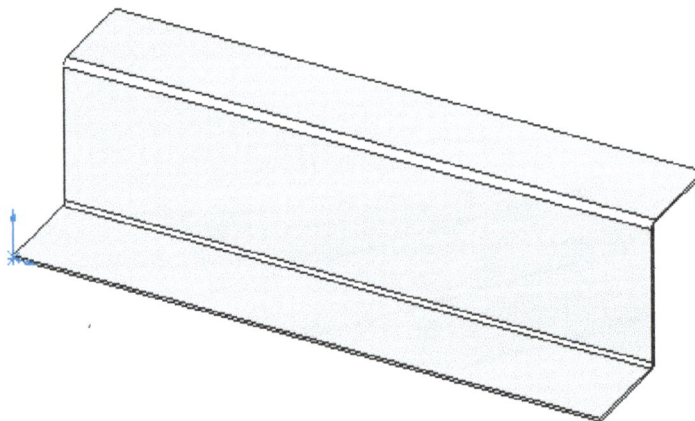

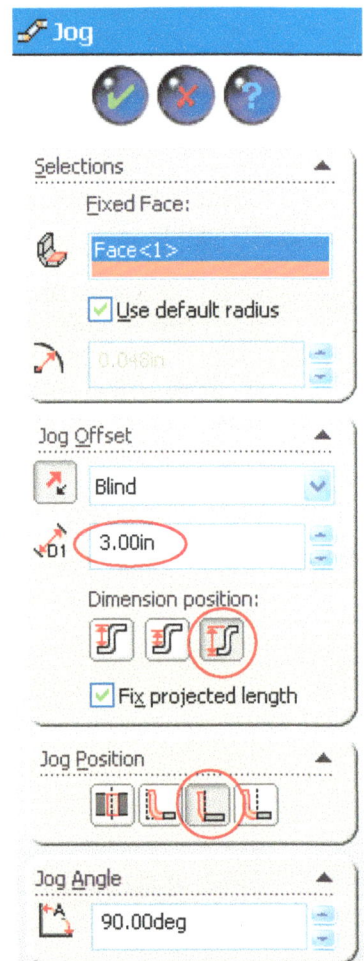

Add Some Holes

In adding holes to the Z-Bracket, you want to make certain that you dimension them so that the holes will be correct even if the other part dimensions are changed.

To do this, create a hole by selecting the **Simple Hole** icon in the CommandManager, or pull down the "Insert" menu and select **Features – Hole – Simple**.

When prompted in the PropertyManager to select a location on a planer face for the center of the hole, pick the top surface of the lower flange of the Z-Bracket.

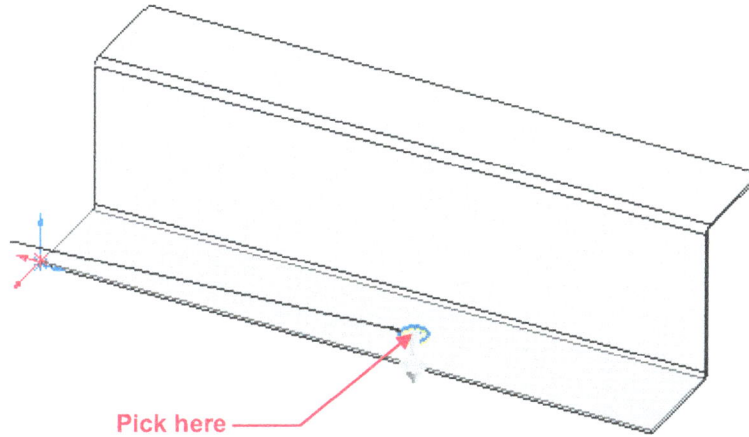

Pick here

In the **Hole** PropertyManager, enter '**0.50in**' for the **Hole Diameter**, and check the **Link to thickness** check box.

Select the green check mark button in the **Hole** PropertyManager to accept the settings and create the hole.

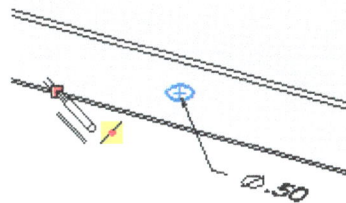

Since the hole is placed at the cursor location, you will need to dimension the sketch for more accuracy.

In the FeatureManager design tree, right click on **Hole1** and select **Edit Sketch**.

Select the **Centerline** icon in the CommandManager, or pull down the "Tools" menu and select **Sketch Entities – Centerline**.

Move the cursor along the lower edge of the part until the icon changes to indicate you are at the midpoint. The line will un-highlight and the cursor will change as shown. Click the left mouse button to select this position.

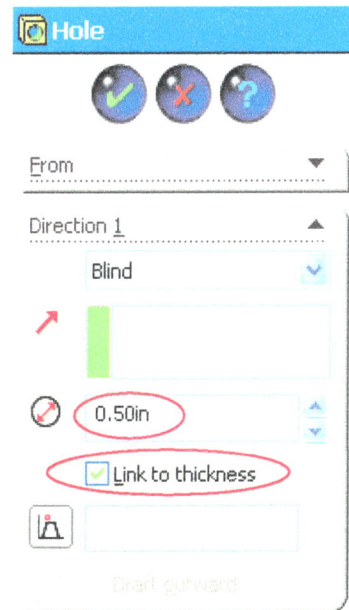

Move the cursor to the other side of the face (flange), again placing the cursor at the midpoint, and click to create a centerline on this flange.

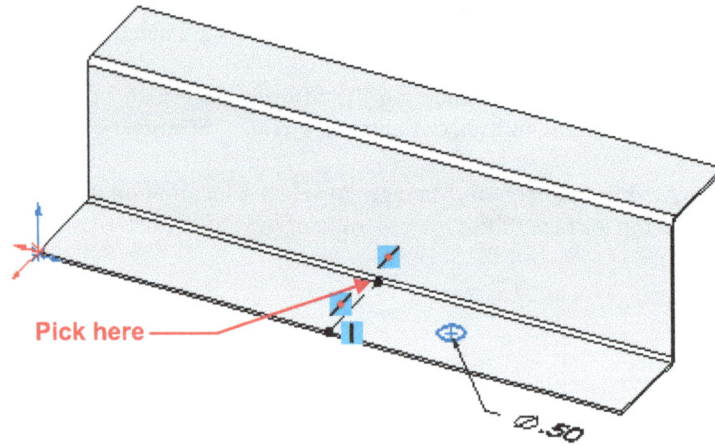

Select the **Smart Dimension** icon in the CommandManager, or pull down the "Tools" menu and select **Dimensions – Smart**.

Select the center point of the circle (a large red dot appears) and the centerline you just created and place a '**2.00**' dimension.

Next, add another dimension from the circle to the vertical flange by selecting the center point of the circle and the top fillet line at the bottom of the vertical flange and place a '**1.00**' dimension.

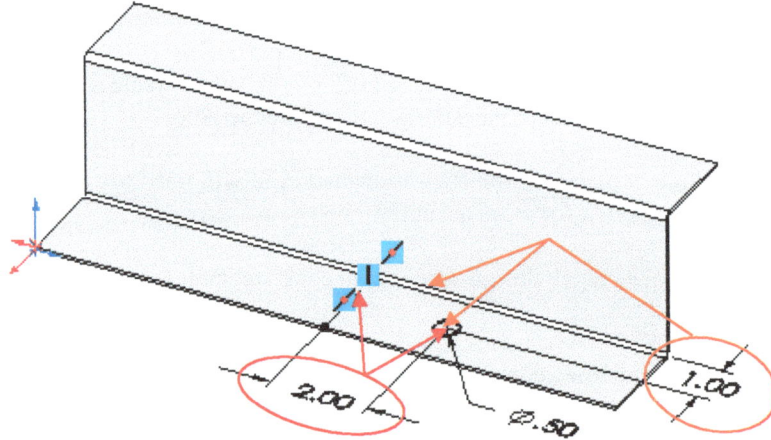

Select the **Mirror Entities** icon in the CommandManager, or pull down the "Tools" menu and select **Sketch Tools - Mirror**.

For the **Entities to mirror**, select the circle (not the center point).

In the **Mirror** PropertyManager, click in the **Mirror about** box to highlight it and then select the centerline you created on the part.

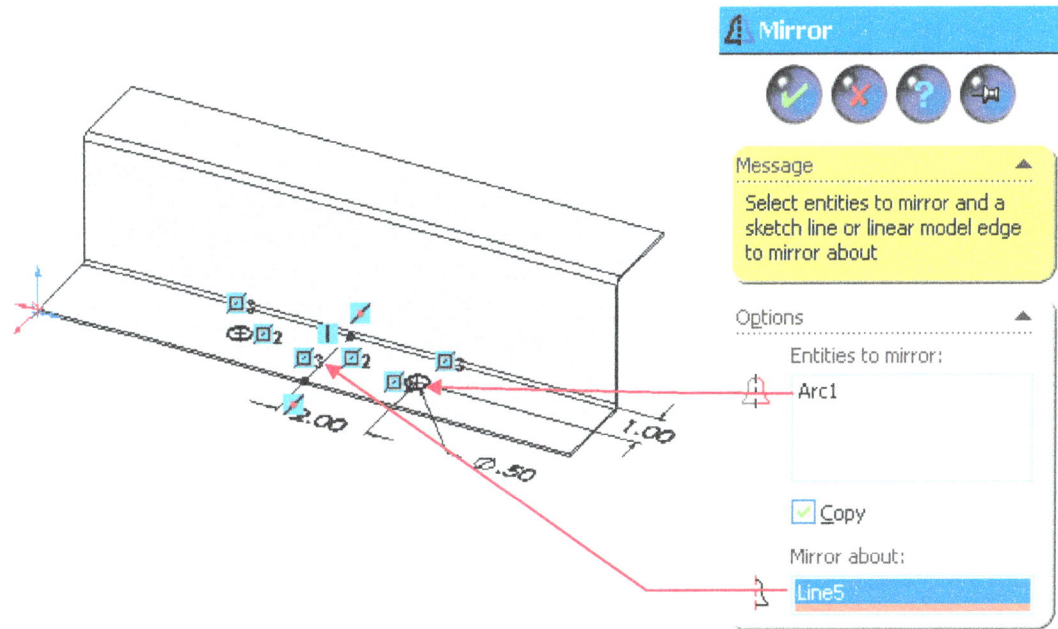

Mirror

Message

Select entities to mirror and a sketch line or linear model edge to mirror about

Options

Entities to mirror:

Arc1

☑ Copy

Mirror about:

Line5

Select the green check mark button at the top of the **Mirror** PropertyManager or press the right mouse button (OK) to mirror the hole.

Exit the sketch by selecting the **Exit Sketch** icon in the CommandManager or in the upper right corner of the graphics area.

Repeat the above steps to add holes to the top flange. These holes are '**0.625in**' diameter, '**5.00**' inches from the centerline, and '**1.25**' inches from the lower fillet line at the top of the vertical flange.

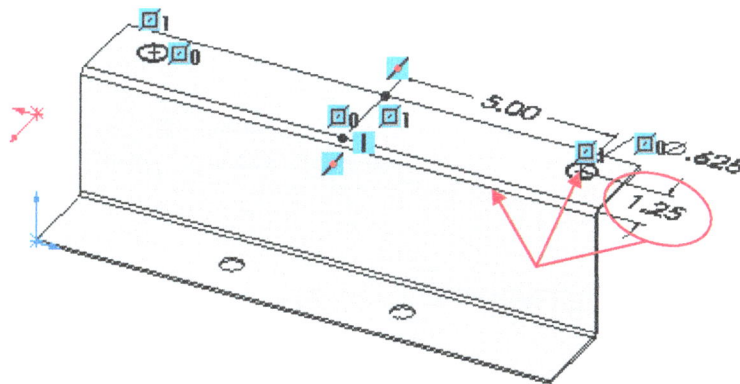

Show the Dimensions

To display the dimensions, in the FeatureManager design tree, right click on **Annotations** and select **Show Feature Dimensions**.

If the part is off the screen, select the **Zoom to Fit** icon in the "View" toolbar, or pull down the "View" menu and select **Modify – Zoom to Fit** so you can see the entire part and center it in the graphics area.

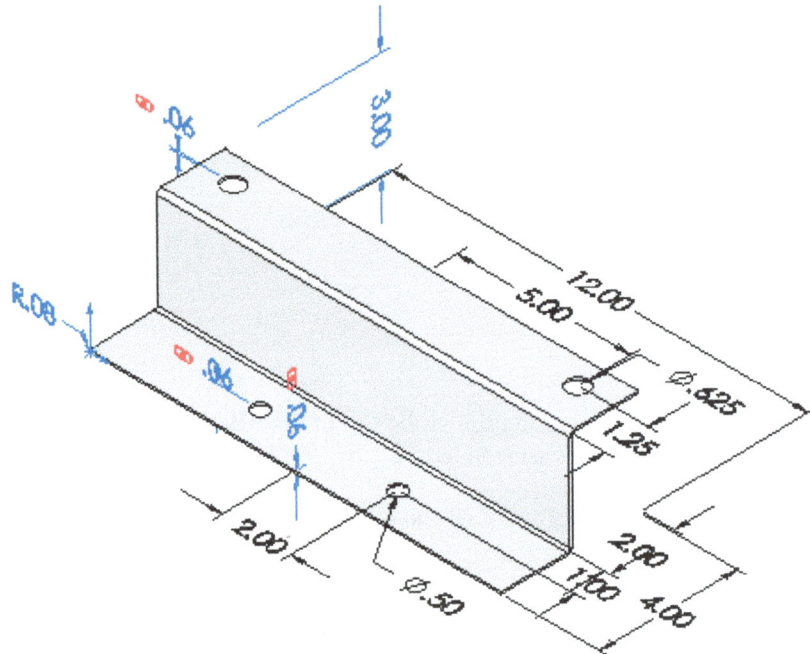

Save the Part

Select the **Save** icon in the "Standard" toolbar, or select **Save** from the "File" pull down menu.

The **Save As** dialog box appears. In the **File name** box, type the name of the drawing number. For this lesson, use '**30636-62**' and select **Save**. The name specified here will be used in the next steps, so please use the name shown.

Using the ConfigurationManager

To activate the ConfigurationManager, simply left click the **ConfigurationManager** tab at the top of the left panel. To return to the FeatureManager design tree, left click the **FeatureManager design tree** tab. SolidWorks automatically creates a default configuration. The configuration is named **Default [Part1]**.

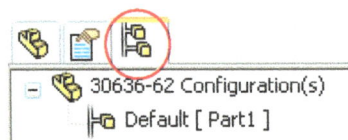

Edit the Configuration Name and Description

In the ConfigurationManager, right click on the **Default** configuration and select **Properties**.

In the **Configuration Properties** PropertyManager, type '**30636-62-1**' for the **Configuration name**. Each configuration must have a unique name. No two configurations can have the same name.

In the **Description** field, type '**16 Ga, H=3**', indicating that this configuration uses 16 gauge material and the overall height is 3 inches.

Select the green check mark button at the top of the **Configuration Properties** PropertyManager. The configuration name changed, but the description is not shown in the tree.

To show the description in the tree, right click the part name, **30636-62 Configuration(s)**, at the top of ConfigurationManager, and select **Tree Display – Show Configuration Descriptions**.

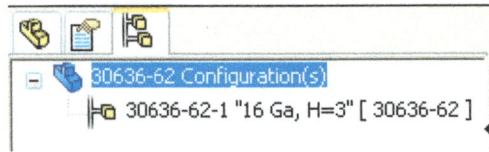

You may need to widen the ConfigurationManager to see the entire description. To do so, drag the right vertical divider bar to the right as shown above.

Create a New Configuration

To create a new configuration, right click on the part name at the top of the ConfigurationManager, **30636-62 Configuration(s)**, and select **Add Configuration**.

In the **Add Configuration** PropertyManager, enter '**30636-62-2**' for the **Configuration Name** and '**14 Ga, H=3**' for the **Description**.

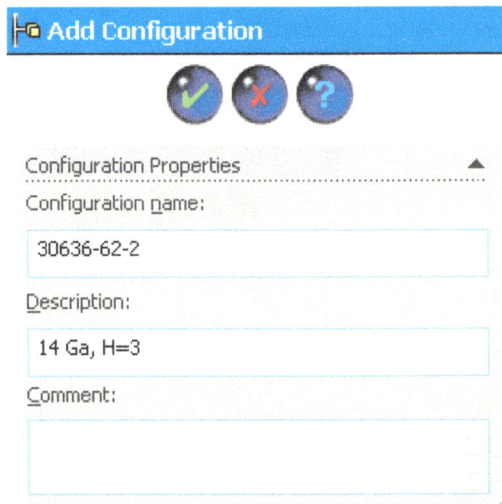

Select the green check mark button at the top of the **Add Configuration** PropertyManager to accept the settings and create the configuration.

The configuration named **30636-62-2 "14 Ga, H=3"** is now active. You can change the dimensions in sketches and in feature definitions and it will only affect the active configuration.

Double click on one of the linked thickness dimensions (**.06**). In the **Modify** dialog box, pull down the menu and select **This Configuration**.

SolidWorks for the Sheet Metal Guy

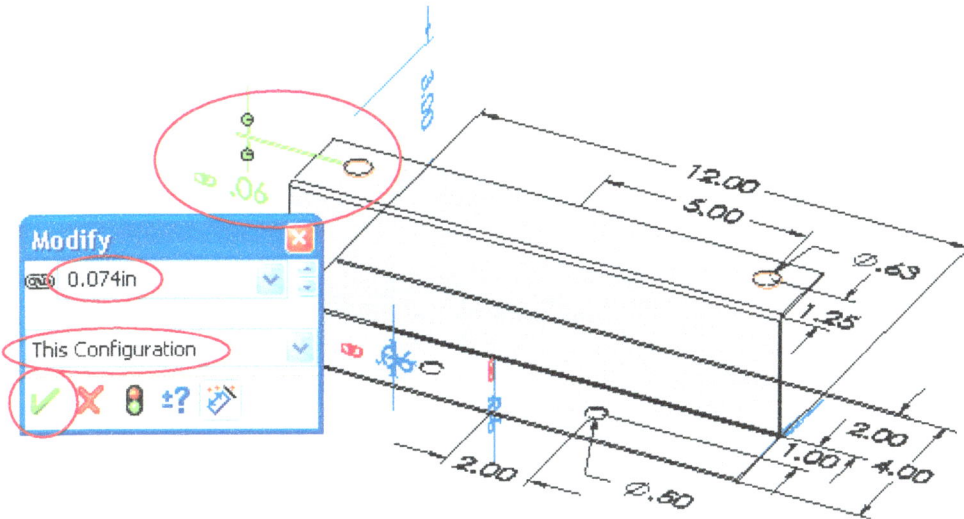

Change the value to '**0.074in**' (14 Gauge).

Select the green check mark button in the **Modify** dialog box.

The linked material thickness value is updated to **.07**. Select the green check mark button at the top of the **Dimension** PropertyManager to accept the settings.

In the ConfigurationManager, double click on the **30636-62-1 "16 Ga, H=3"** configuration name to make it active. (The text will change from grey to black). A warning dialog box may appear indicating that the part needs updating. Check **Don't ask me again** and then select **Yes** to rebuild the model. The material thickness dimension updates back to the original **.06** (16 Gauge) dimension.

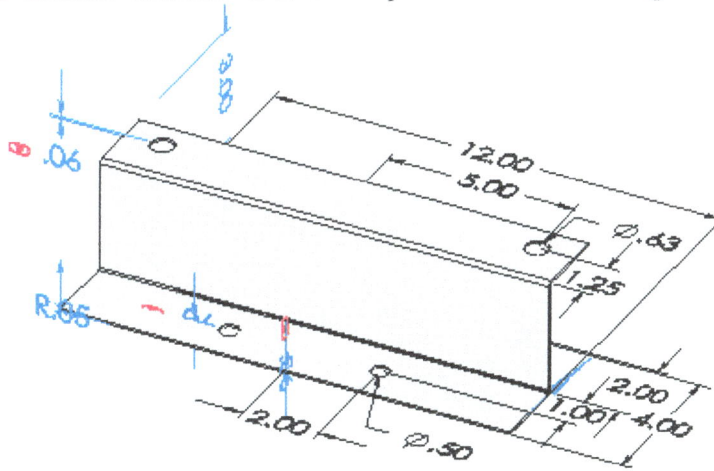

There are two ways you can tell which configuration is active. First, the active configuration is black, the same color as the top line of the ConfigurationManager. The other configuration names are gray. The second way of knowing is to look at the top line. The active configuration is listed here to the right. You may need to widen the ConfigurationManager window to see this.

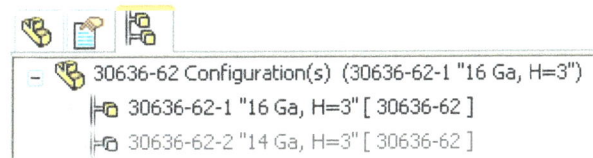

Create a New Configuration with a Different Height

With Configuration **30636-62-1 "16 Ga, H=3"** active, right click on the part name at the top of the ConfigurationManager, and select **Add Configuration**.

In the **Add Configuration** PropertyManager, enter '**30636-62-3**' for the **Configuration Name** and '**16 Ga, H=2.25**' for the **Description**.

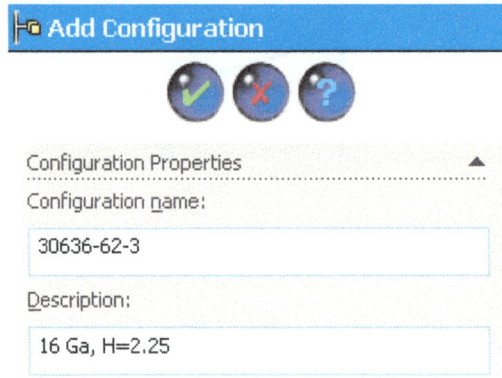

Select the green check mark button at the top of the **Add Configuration** PropertyManager to accept the settings and create the configuration.

The configuration named **30636-62-3 "16 Ga, H=2.25"** is now active.

Find the **3.00** dimension for the height of the **Jog Offset**. Double click this **3.00** dimension and change the value to '**2.25in**'.

Pull down the menu immediately below the value and select **This Configuration**.

In the **Modify** dialog box, select the **Rebuild** icon to regenerate the model with the current value.

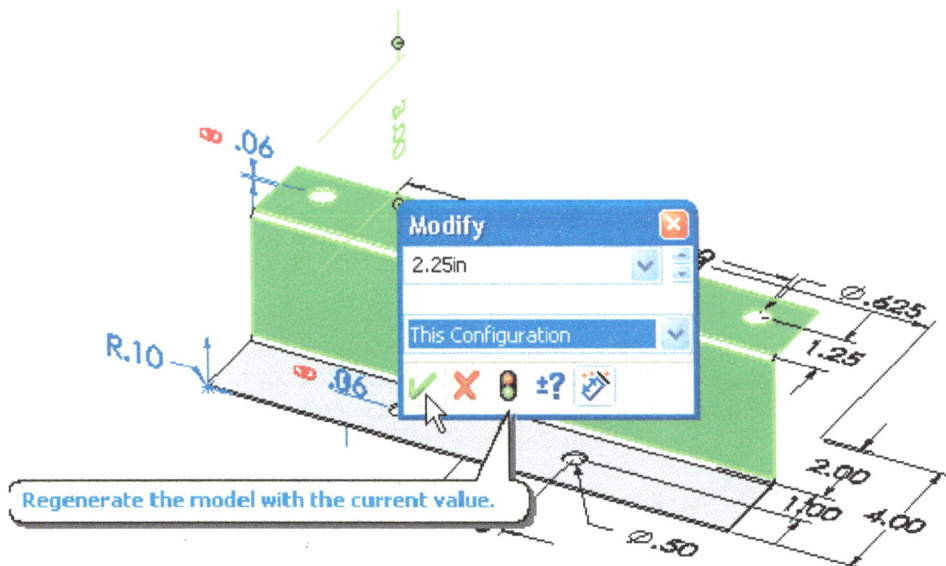

Regenerate the model with the current value.

Then, in the **Modify** dialog box, select the green check mark icon to save the current value and exit the dialog.

Select the green check mark button at the top of the **Dimension** PropertyManager.

Once you have created multiple configurations, use caution when editing a value through the FeatureManager design tree as you cannot tell whether the new value is for **This Configuration** or **All Configurations**. It is recommended that you edit the dimension value in the graphics area where the dialog box will show you whether it is for **This Configuration** or **All Configurations**.

Since the **30636-62-3 "16 Ga, H=2.25"** configuration was created with the **30636-62-1 "16 Ga, H=3"** configuration active, the **30636-62-3 "16 Ga, H=2.25"** configuration has a 16 gauge material thickness. If the 14 gauge configuration was active when you created **30636-62-3**, the material thickness would be 14 gauge.

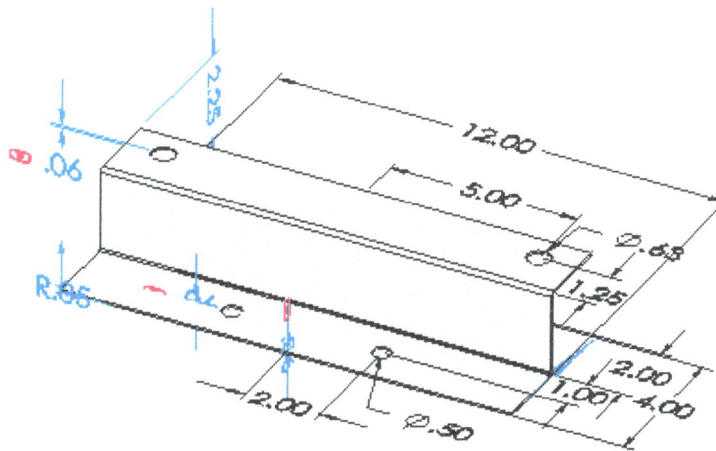

Create a New Configuration with Multiple Changes

Click the **ConfigurationManager** tab at the top of the left panel, right click on the part name at the top of the ConfigurationManager, and select **Add Configuration**.

In the **Add Configuration** PropertyManager, enter '**30636-62-4**' for the **Configuration Name** and '**12 Ga, H=5, L=14**' for the **Description**.

Select the green check mark button at the top of the **Add Configuration** PropertyManager to accept the settings and create the configuration.

In the graphics area, find the material thickness dimension and double click on it. Enter the value '**0.1046in**'. Make certain the pull down window below the value says **This Configuration**, and then select the green check mark icon in the **Modify** dialog box or press the **Enter** key to accept the change.

Find the Jog Height dimension. Double click on it and in the **Modify** dialog box, change the dimension to '**5.00in**'. Pull down the menu and select **This Configuration** and press the **Enter** key to accept the change.

Double click on the **12.00** overall length dimension.

In the **Modify** dialog box, change the dimension to '**14.00in**'.

Pull down the menu and select **This Configuration** and press the **Enter** key to accept the change.

Select the **Rebuild** icon in the "Standard" toolbar, or press **Ctrl-B**.

You may click and drag the dimension to a new location at any time.

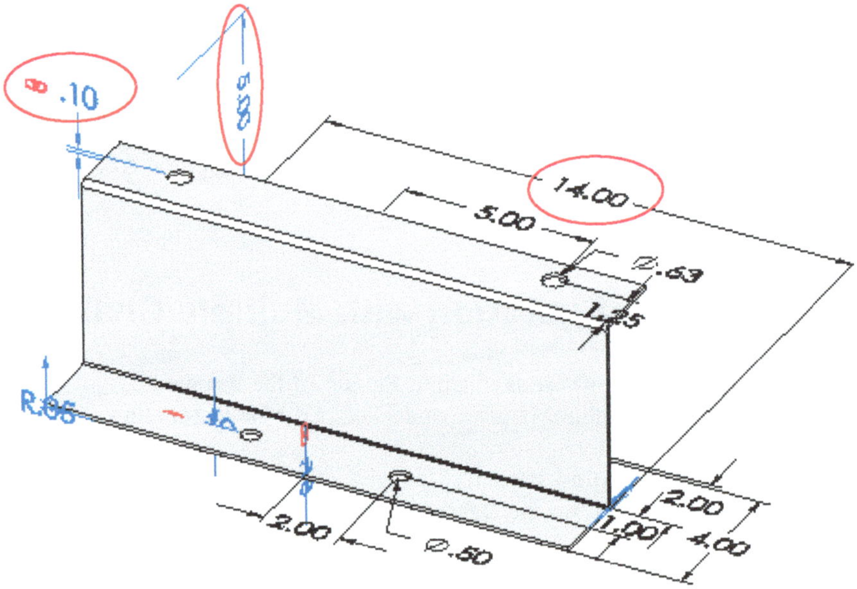

SolidWorks for the Sheet Metal Guy

Configuration with Suppressed Features

Click the **ConfigurationManager** tab at the top of the left panel, right click on the part name at the top of the ConfigurationManager, and select **Add Configuration**.

In the **Add Configuration** PropertyManager, enter '**30636-62-5**' for the **Configuration Name** and '**12 Ga, H=5, L=14, No Holes**' for the **Description**.

Select the green check mark button at the top of the **Add Configuration** PropertyManager to accept the settings and create the configuration.

Left click the **FeatureManager design tree** tab to return to the FeatureManager design tree and right click on **Hole1** and select **Suppress**. The hole feature is removed from the current configuration only.

Right click on **Hole2** and select **Suppress**. The hole feature is removed from the current configuration only.

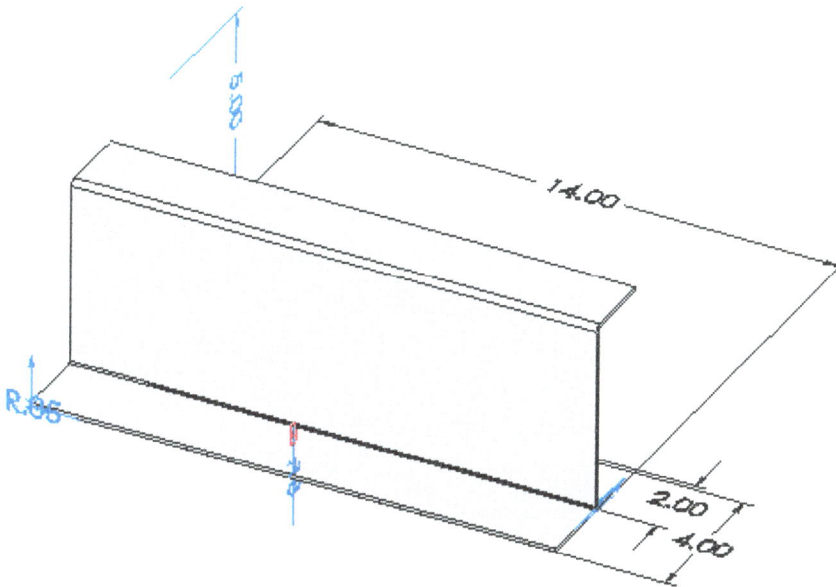

The final ConfigurationManager is shown below.

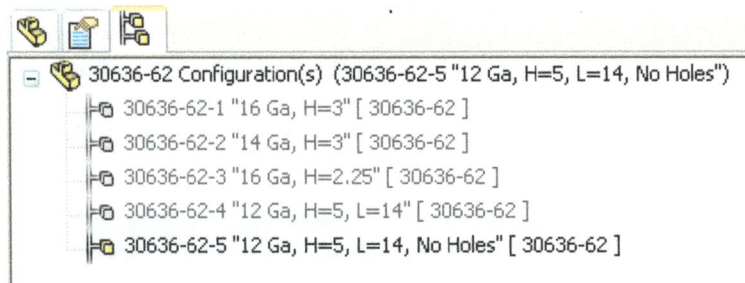

Save the Part

Select the **Save** icon in the "Standard" toolbar, or select **Save** from the "File" pull down menu.

Since the part was already saved once, SolidWorks replaces the original with the updated version.

Chapter 10

Rack Panel

The Rack Panel puts it all together. It utilizes many of the command features you have already used and adds a few new twists. At one point, you will create a tab on the end flange and then a mating tab on the side flange. A single hole command is used to place the same hole in both tabs.

To control the length of the hem at the end of the part, you must create a notch.

Again a tab is added using the Base Flange/Tab command. And again, this could have been done by modifying the sketch of the original flange.

Create the Base Flange

Begin a new part document by selecting the **New** icon in the "Standard" toolbar, or pull down the "File" menu and select **New**.

Select **Part** in the **New SolidWorks Document** dialog box and then select **OK**.

Create a base flange by selecting the **Sheet Metal** icon in the control area of the CommandManager. Then, select the **Base-Flange/Tab** icon from the toolbar, or pull down the "Insert" menu and select **Sheet Metal – Base Flange**.

Select the **Top** plane when prompted to select a plane on which to sketch the feature cross-section.

Create a rectangle with the lower left hand corner starting at the origin using the **Rectangle** icon in the CommandManager, or pull down the "Tools" menu and select **Sketch Entities – Rectangle**.

Add a '**30.00**' horizontal dimension to the bottom line and a '**20.00**' vertical dimension to the left vertical line using the **Smart Dimension** icon in the CommandManager, or pull down the "Tools" menu and select **Dimensions – Smart**.

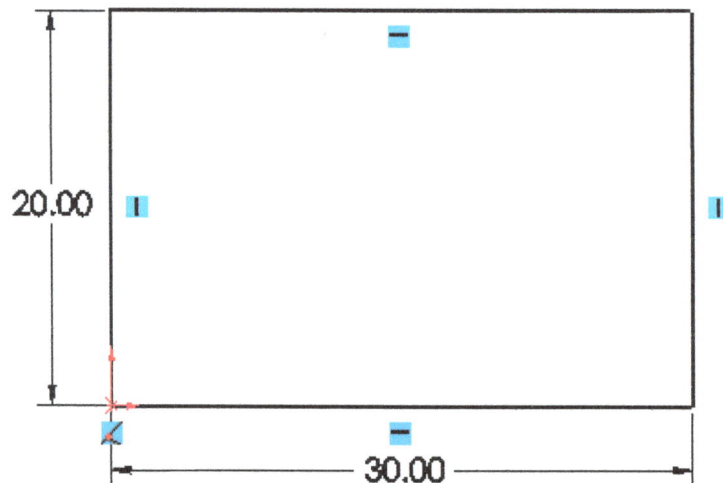

Exit the sketch by selecting the **Exit Sketch** icon in the CommandManager or in the upper right corner of the graphics area. You don't have to use the **Zoom to Fit** command before exiting the sketch. SolidWorks will do this automatically when it shows the preview of the Base Flange.

In the **Base Flange** PropertyManager under **Sheet Metal Parameters**, set the **Thickness** to '**0.048in**'.

Check the **Reverse direction** check box to make sure that the preview of the extrusion is in the Y-positive direction.

Select the green check mark button at the top of the **Base Flange** PropertyManager to accept the settings and create the feature.

Add an Edge Flange

Select the **Edge Flange** icon in the CommandManager, or pull down the "Insert" menu and select **Sheet Metal – Edge Flange**.

Select the back left edge and move the cursor up and click to set the direction of the flange.

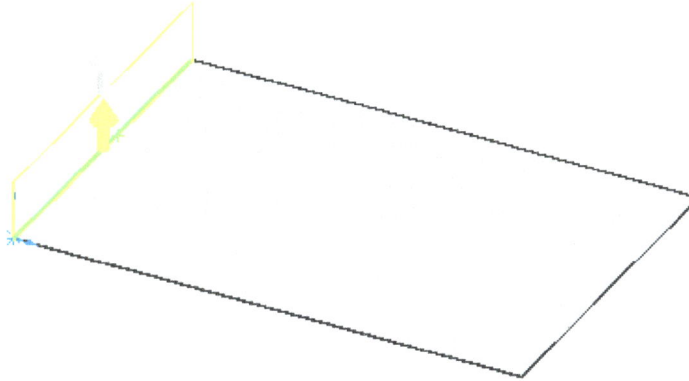

In the **Edge-Flange** PropertyManager, set the **Flange Length** to **Blind**, and the **Length** to '**1.00in**'.

Select the **Outer Virtual Sharp** button and the **Material Inside Flange Position** button.

Select the green check mark button at the top of the **Edge-Flange** PropertyManager to accept the settings and create the flange.

In the FeatureManager design tree, right click on the **Edge-Flange1** feature and select **Edit Sketch** from the menu.

Change the display view to **Normal To** the sketch by selecting **Normal To** in the View list in the lower left corner of the graphics area.

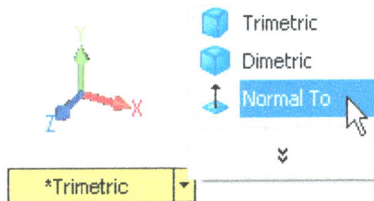

Left click and drag the vertical edge lines in to the middle.

Then, select the **Smart Dimension** icon and add a '**3.00**' dimension to the left side as shown on the next page.

On the right side, pick the two points to create the dimension. When the **Modify** dialog box opens, pull down the menu and select **Add Equation**. Click on the **3.00** dimension on the left side of the part to set the two dimensions equal to one another. You may have to drag the Equation dialog box out of the way to access the dimension. Then, select the **OK** button to close the **Add Equation** dialog box. Then select the **OK** button in the **Equations** dialog box.

Exit the sketch by selecting the **Exit Sketch** icon in the CommandManager or in the upper right corner of the graphics area.

Change the display view to **Trimetric** by selecting **Trimetric** in the View list in the lower left corner of the graphics area.

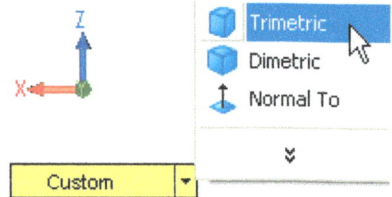

Add a Small Tab

Select the **Edge Flange** icon in the CommandManager, or pull down the "Insert" menu and select **Sheet Metal – Edge Flange**.

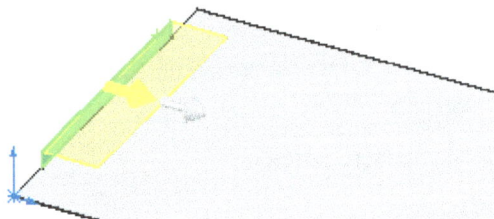

Select the top edge of the flange you just created, and move the cursor to the right and click to set the direction of the flange.

In the **Edge-Flange** PropertyManager, set the **Flange Length** to **Blind** and the **Length** to '1.00in'. Select the **Outer Virtual Sharp** button and the **Material Inside Flange Position** button.

Next, select the **Edit Flange Profile** button.
| Edit Flange Profile |

Left click and drag the left edge line towards the middle.

Then, select the **Smart Dimension** icon and add a '1.00' dimension as shown to the right.

Select the **Finish** button in the **Profile Sketch** dialog box.
| Finish |

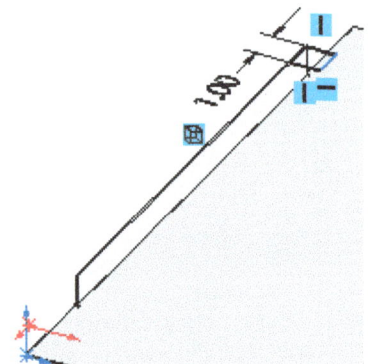

Create a Miter Flange

Select the **Miter Flange** icon in the CommandManager, or pull down the "Insert" menu and select **Sheet Metal – Miter Flange**.

Change the display view to **Left** by selecting **Left** in the View list in the lower left corner of the graphics area.

You are prompted in the PropertyManager to select a plane, a planar face or an edge on which to sketch the feature cross-section. Select the face of the edge flange as shown.

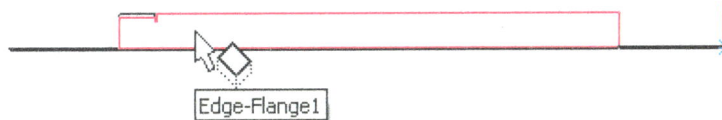

Using the **Line** tool, create a '**3.00**' vertical line starting at the bottom left corner of the part. Make sure that the line starts at the bottom line, not the top line, to control the overall part height. Use the **Zoom to Area** command if you need to in order to select the correct locations.

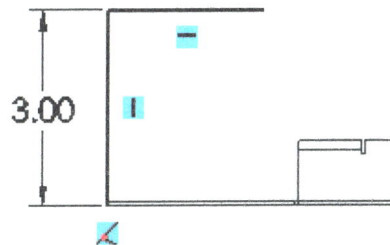

Next, create a horizontal line to the right of the top of the vertical line.

Select the **Add Relation** icon in the CommandManager, or pull down the "Tools" menu and select **Relations – Add**.

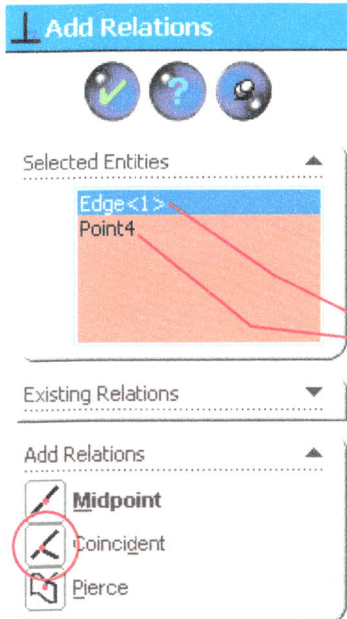

Select the left edge (not the top point) of the edge flange and the right endpoint of the horizontal line as shown.

In the **Add Relations** PropertyManager, select the **Coincident** relation button.

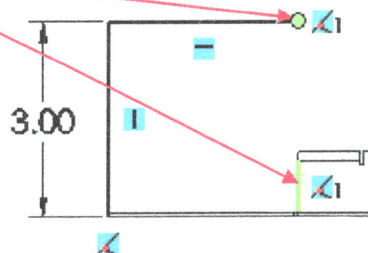

Select the green check mark button at the top of the **Add Relations** PropertyManager to accept the settings.

Exit the sketch by selecting the **Exit Sketch** icon in the CommandManager or in the upper right corner of the graphics area.

Make sure that the **Flange position** is set to **Material Inside**.

Select the green check mark button at the top of the **Miter Flange** Property-Manager to accept the settings and create the miter flange.

Change the display view to **Trimetric** by selecting **Trimetric** in the View list in the lower left corner of the graphics area.

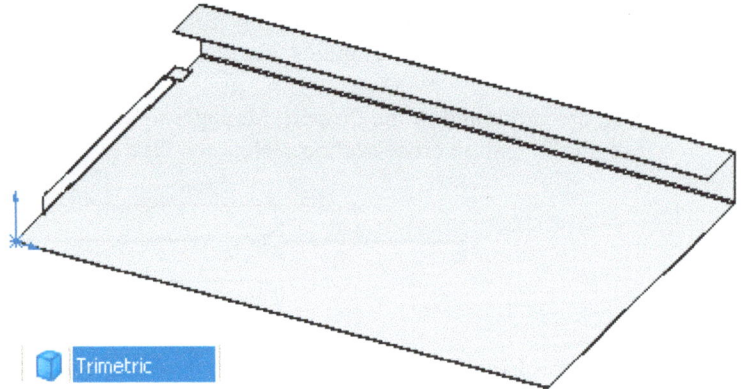

Add to the Flange

Select the **Edge Flange** icon in the CommandManager, or pull down the "Insert" menu and select **Sheet Metal – Edge Flange**.

Select the front edge of the miter flange and move the cursor down and click to set the direction of the flange.

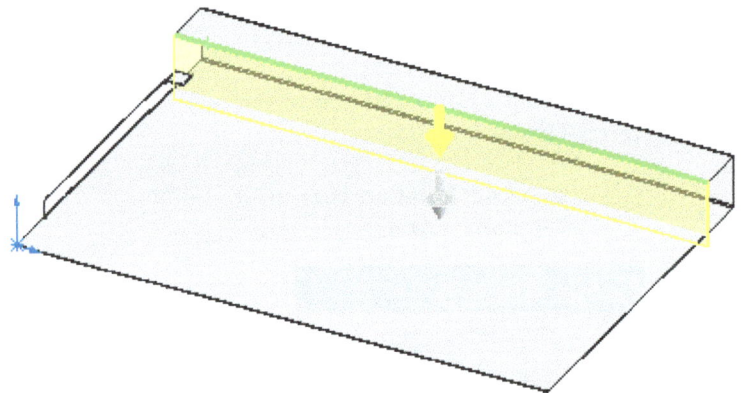

In the **Edge-Flange** PropertyManager, select the **Edit Flange Profile** button.

Change the display view to **Normal To** the sketch by selecting **Normal To** in the View list in the lower left corner of the graphics area.

Left click and drag the right edge line to the far left as shown.

Using the **Line** tool, create a vertical and a horizontal line.

Select the **Trim** icon in the CommandManager, or pull down the "Tools" menu and select **Sketch Tools – Trim**.

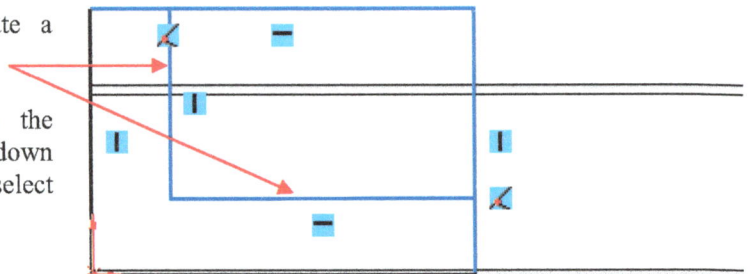

Select the top portion of the right vertical line that you want to trim, and select the right portion of the top horizontal line that you want to trim.

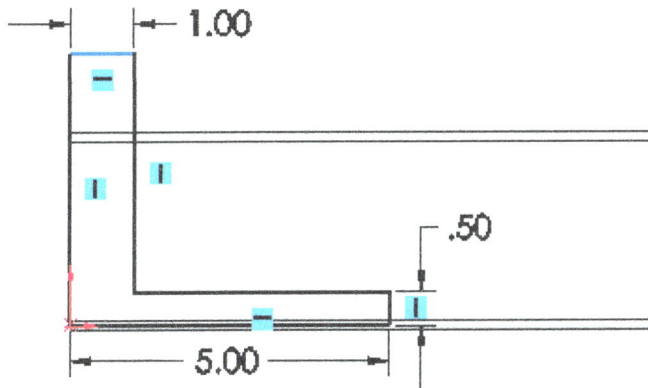

Select the **Smart Dimension** icon and add the following dimensions.

Select the **Back** button in the **Profile Sketch** dialog box.

| < Back |

Change the display view to **Trimetric** by selecting **Trimetric** in the View list in the lower left corner of the graphics area.

In the **Edge-Flange** PropertyManager, set the **Flange Length** to '**2.00in**'. Select the **Outer Virtual Sharp** button. In **Flange Position**, select the **Material Inside** button.

Select the green check mark button at the top of the **Edge-Flange** PropertyManager to accept the settings and create the feature.

The **What's Wrong** dialog box appears.

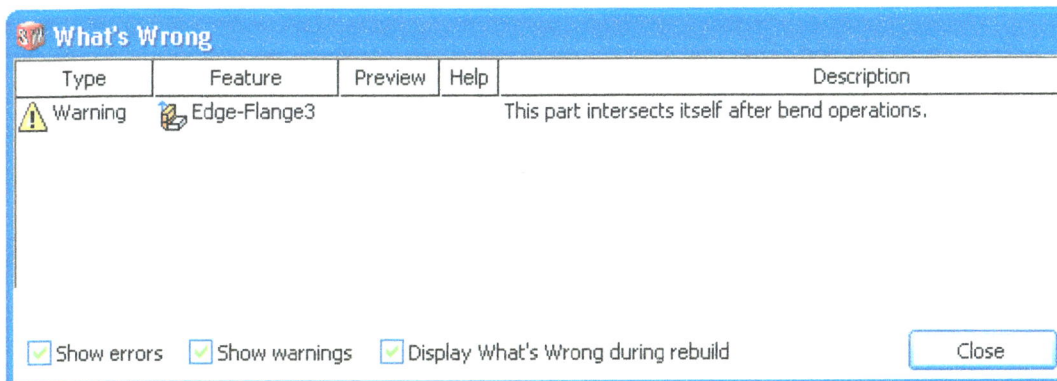

Type	Feature	Preview	Help	Description
⚠ Warning	Edge-Flange3			This part intersects itself after bend operations.

☑ Show errors ☑ Show warnings ☑ Display What's Wrong during rebuild | Close |

Since the overall height is **3.00** and the side flange height is **1.00**, SolidWorks cannot handle the **2.00** dimension. To fix the problem, you need to shorten the length of the flange by a small amount, let's say **.01** to ensure that the surfaces do not touch.

Right click **Edge-Flange3** and select **Edit Feature**.

In the **Edge-Flange** PropertyManager, set the **Flange Length** to '**1.99in**'.

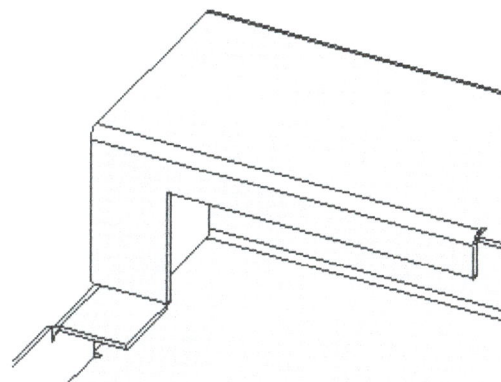

Select the green check mark button at the top of the **Edge-Flange** PropertyManager to accept the settings and update the feature. The error message is now gone.

Creating the Tab

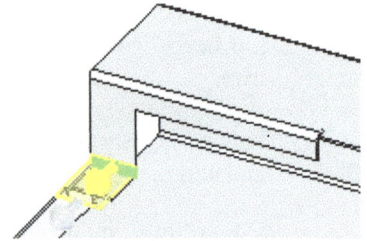

Select the **Edge Flange** icon in the CommandManager, or pull down the "Insert" menu and select **Sheet Metal – Edge Flange**.

Select the bottom edge of the flange you just created and move the cursor down and to the left and click to set the direction of the flange. **Zoom to area** if you need to get a better view.

In the **Edge-Flange** PropertyManager, select the **Edit Flange Profile** button.

> Edit Flange Profile

Select the bottom horizontal line. Next, hold down the **Control** key (**Ctrl**) on the keyboard. Then, select the bottom edge line of the small tab as shown.

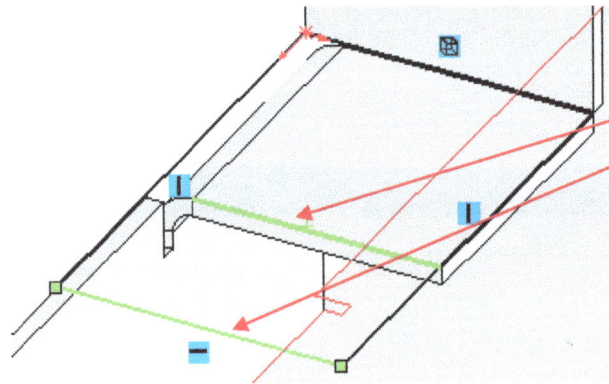

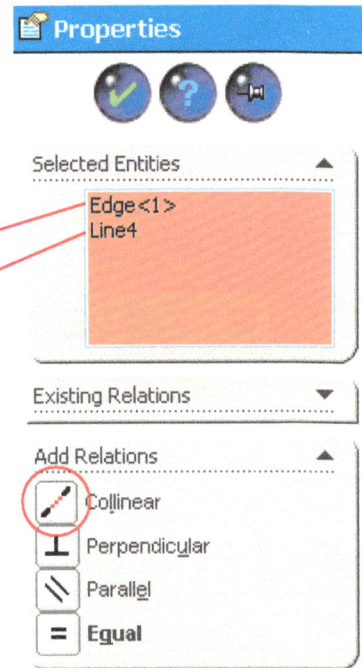

In the **Properties** PropertyManager, select the **Collinear** relation button.

Select the green check mark button at the top of the **Properties** PropertyManager.

Select the **Finish** button in the **Profile Sketch** dialog box.   > Finish

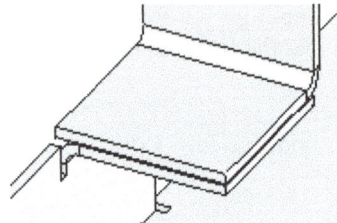

Create a Hole Through the Tabs

Select the **Simple Hole** icon in the CommandManager, or pull down the "Insert" menu and select **Features – Hole – Simple**.

When prompted in the PropertyManager to select a location on a planer face for the center of the hole, pick the top surface of the small flange you just created.

In the **Hole** PropertyManager, pull down the menu and change the **End Condition** to **Up To Surface**.

Move the cursor into the graphics area and right click on the edge of the lower tab flange. From the menu, select **Select Other**.

The **Select Other** dialog box opens, offering you several choices. As you move the cursor over the different choices, the geometry representing the item highlights. Select the bottom face of the lower tab as shown.

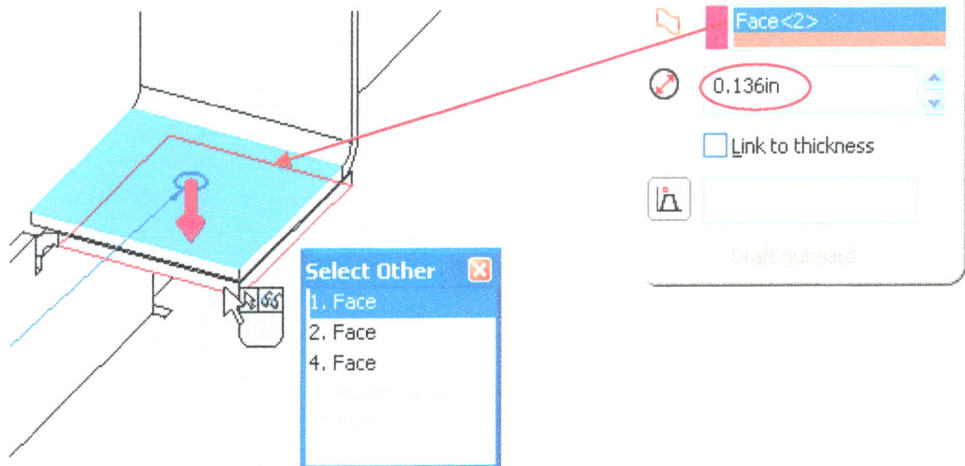

Enter '**0.136in**' for the **Hole Diameter**.

Select the green check mark button to accept the settings and create the hole.

In the FeatureManager design tree, right click on **Hole1** and select **Edit Sketch**.

Select the **Centerline** icon in the CommandManager, or pull down the "Tools" menu and select **Sketch Entities – Centerline**.

Create a vertical line connecting the two midpoints of the flange as shown. Make certain the red dots appear, indicating that you are picking the midpoints.

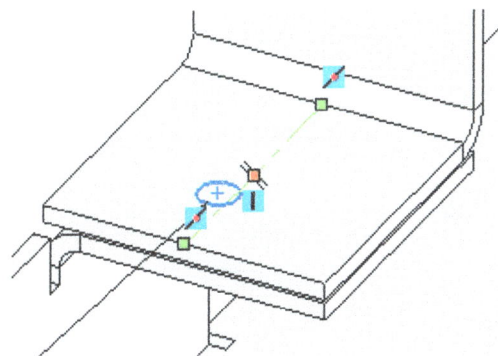

Select the **Centerline** icon in the CommandManager, or pull down the "Tools" menu and select **Sketch Entities – Centerline** to end the command.

With the cursor, left click and drag the center point of the circle to the midpoint of the centerline. The red dot appears at the center when the cursor is there. A relation will automatically be created, locking in the circle at the midpoint of the centerline.

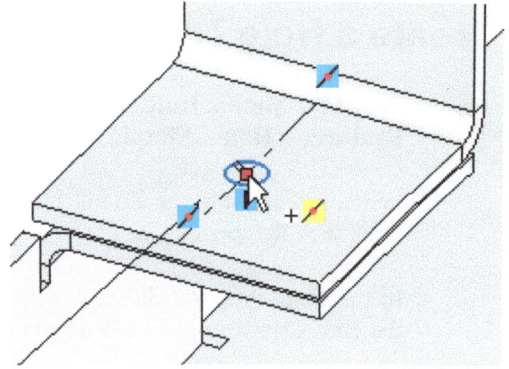

Exit the sketch by selecting the **Exit Sketch** icon in the CommandManager or in the upper right corner of the graphics area.

Add Another Edge Flange

Select the **Edge Flange** icon in the CommandManager, or pull down the "Insert" menu and select **Sheet Metal – Edge Flange**.

Press the **F** key on the keyboard to **Zoom to Fit** so you can see the entire rectangle and center it in the graphics area. (Keyboard shortcut key, **Zoom to Fit**: f)

Select the top edge of the miter flange and move the cursor down and click to set the flange direction.

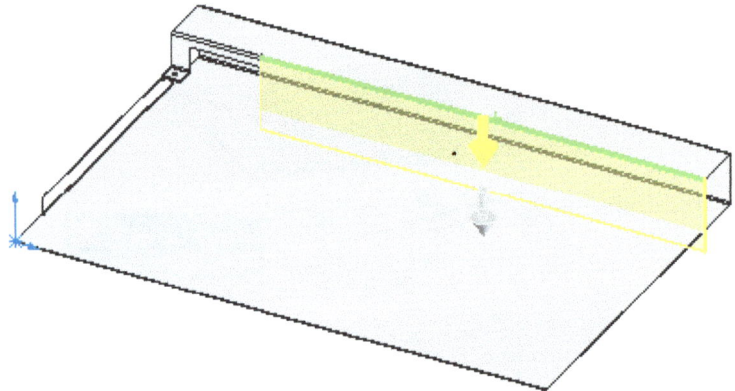

In the **Edge-Flange** PropertyManager, select the **Edit Flange Profile** button.   Edit Flange Profile

Left click and drag the left vertical line to the right.

Select the **Smart Dimension** icon and add a '**7.00**' dimension.

Select the **Back** button in the **Profile Sketch** dialog box.   < Back

In **Flange Position**, set the **Flange Length** to '**0.50in**' and select the **Material Inside** button.

Select the green check mark button at the top of the **Edge-Flange** PropertyManager.

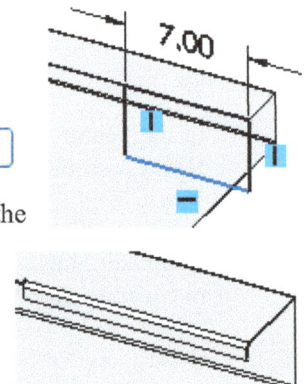

Create Another Flange

Select the **Edge Flange** icon in the CommandManager, or pull down the "Insert" menu and select **Sheet Metal – Edge Flange**.

Select the bottom edge of the flange just created. Move the cursor to the left and click to set the direction of the flange.

In the **Edge-Flange** PropertyManager, set the **Flange Length** to '**0.50in**'.

Select the **Bend Outside Flange Position** button.

In the **Edge-Flange** PropertyManager, select the **Edit Flange Profile** button. | Edit Flange Profile |

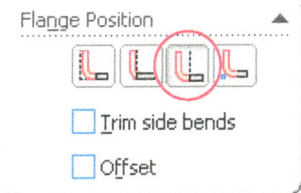

Use the **Zoom To Area** command to enlarge the sketch of the new flange.

Left click and drag the right vertical line to the left, and then select the **Smart Dimension** icon and add a '**5.00**' dimension as shown.

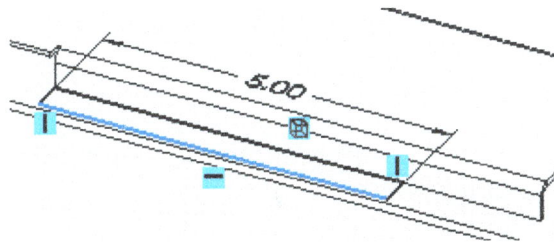

Select the **Finish** button in the **Profile Sketch** dialog box. | Finish |

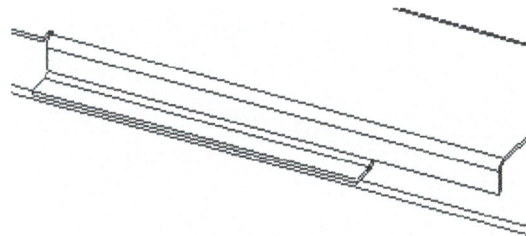

Creating the Middle Flange

Select the **Edge Flange** icon in the CommandManager, or pull down the "Insert" menu and select **Sheet Metal – Edge Flange**.

Press the **F** key on the keyboard to **Zoom to Fit** so you can see the entire rectangle and center it in the graphics area. (Keyboard shortcut key, **Zoom to Fit**: f)

Select the front edge of the miter flange and move the cursor up and click to set the flange direction.

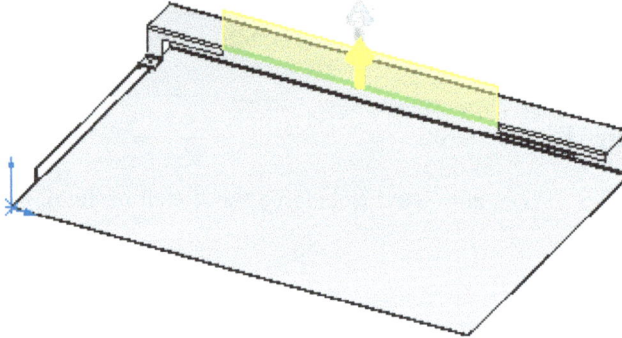

In the **Edge-Flange** PropertyManager, set the **Flange Length** to '**1.00in**'.

Select the **Outer Virtual Sharp** button and the **Material Inside** button.

Select the green check mark button at the top of the **Edge-Flange** PropertyManager to accept the settings and create the flange.

Create a Miter Flange

Select the **Miter Flange** icon in the CommandManager, or pull down the "Insert" menu and select **Sheet Metal – Miter Flange**.

Change the display view to **Left** by selecting **Left** in the View list in the lower left corner of the graphics area.

You are prompted in the PropertyManager to select a plane, a planar face or an edge on which to sketch the feature cross-section. Select the face of the edge flange as shown.

Edge-Flange1

Using the **Line** tool, create a vertical line starting at the bottom right corner of the part. Make sure that the line starts at the bottom line, not the top line, to control the overall part height. Use the **Zoom to Area** command if you need to in order to select the correct location.

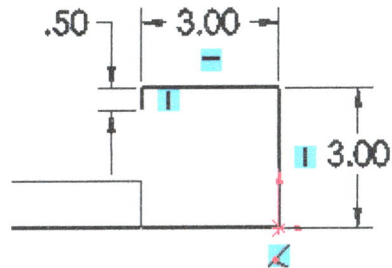

Next, create a horizontal line to the left of the top of the vertical line, and then create another vertical line down from the left endpoint of the horizontal line.

Select the **Smart Dimension** icon and dimension the sketch as shown.

Exit the sketch by selecting the **Exit Sketch** icon in the CommandManager or in the upper right corner of the graphics area.

Make sure that the **Flange position** is set to **Material Inside**.

Select the green check mark button at the top of the **Miter Flange** PropertyManager to accept the settings and create the miter flange.

Change the display view to **Trimetric** by selecting **Trimetric** in the View list in the lower left corner of the graphics area.

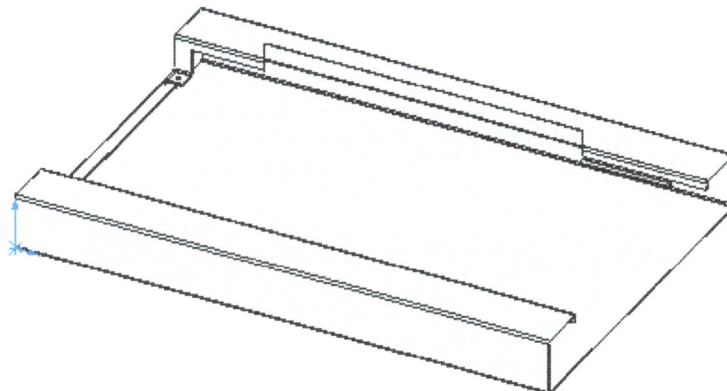

Add a Flange to the Right End

Select the **Edge Flange** icon in the CommandManager, or pull down the "Insert" menu and select **Sheet Metal – Edge Flange**.

Select the right end of the base flange and move the cursor up and click to set the direction of the flange.

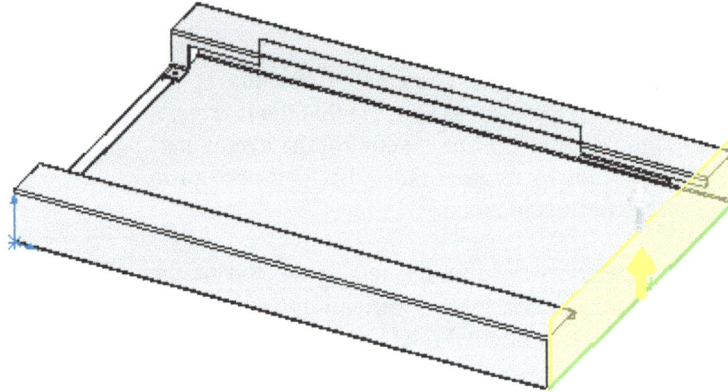

In the **Edge-Flange** PropertyManager, select the **Edit Flange Profile** button. `Edit Flange Profile`

Change the display view to **Normal To** the sketch by selecting **Normal To** in the View list in the lower left corner of the graphics area. `Normal To`

Left click and drag the vertical side lines in towards the middle.

Select the **Smart Dimension** icon and add two '**2.50**' dimensions as shown. The lines are dimensioned to the outside edge line of the part.

Select the **Back** button in the **Profile Sketch** dialog box. `< Back`

In the **Edge-Flange** PropertyManager, set the **Flange Length** to '**1.00in**'.

Select the **Outer Virtual Sharp** button and the **Material Inside** button.

Check the **Custom Relief Type** check box.

Set the **Relief Type** to **Rectangle**.

Uncheck the **Use relief ratio** checkbox and enter '**0.25in**' for the **Relief Width** and '**0.00in**' for the **Relief Depth**.

Select the green check mark button at the top of the **Edge-Flange** PropertyManager to accept the settings and create the flange.

Change the display view to **Trimetric** by selecting **Trimetric** in the View list in the lower left corner of the graphics area.

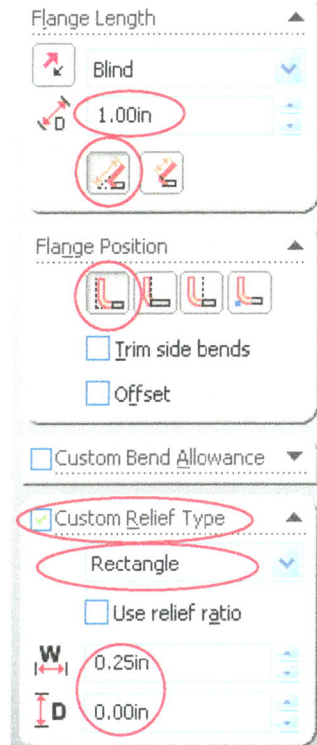

Add a Hem

To control the length of the hem, you must first add a notch at a specific location.

To do this, select the **Extruded Cut** icon in the CommandManager, or pull down the "Insert" menu and select **Cut – Extrude**.

Select the top of the base flange for the sketch plane.

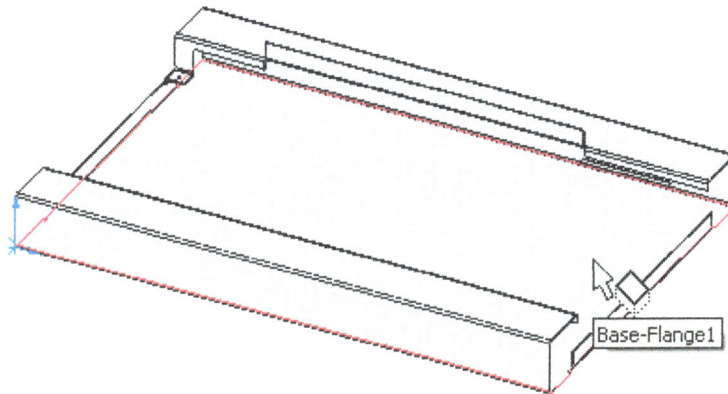

Change the display view to **Bottom** by selecting **Bottom** in the View list in the lower left corner of the graphics area.

In the upper right hand corner, create a rectangle on the right edge line of the part using the **Rectangle** icon in the CommandManager, or pull down the "Tools" menu and select **Sketch Entities – Rectangle**.

Select the **Smart Dimension** icon and add a '**0.25**' and a '**1.00**' vertical dimension to the rectangle as shown.

Add a horizontal dimension. When the **Modify** dialog box appears, select **Add Equation** from the pull down menu.

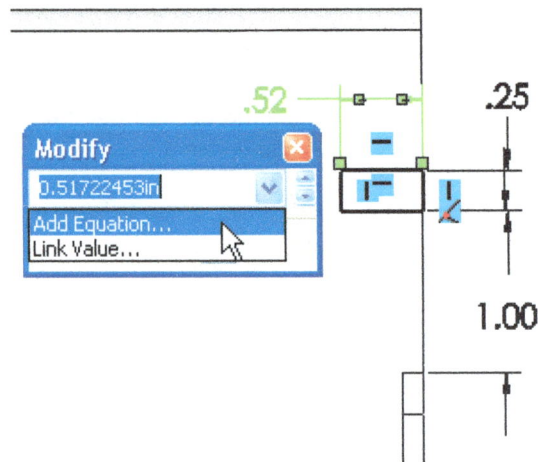

Type in '**"D1@Sheet-Metal1" + Thickness**' in the **Add Equation** dialog box.

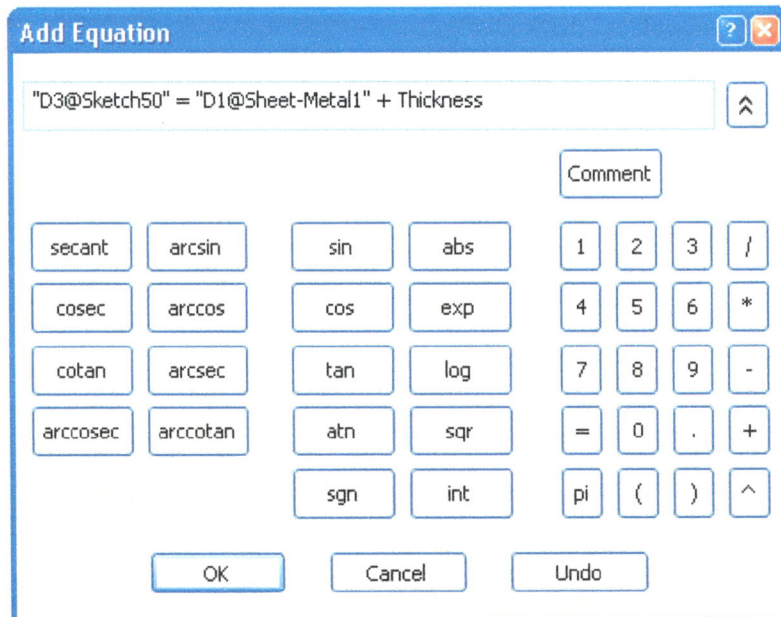

Select the **OK** button in the **Add Equation** dialog box and in the **Equations** dialog box.

SolidWorks for the Sheet Metal Guy

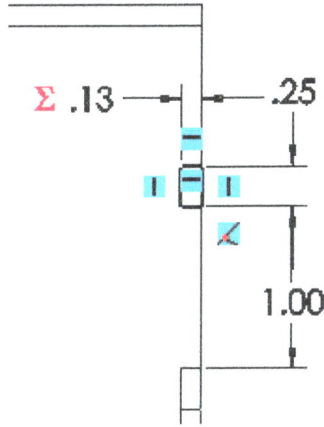

Exit the sketch by selecting the **Exit Sketch** icon in the CommandManager or in the upper right corner of the graphics area.

In the **Cut-Extrude** Property Manager, check the **Link to thickness** check box.

Select the green check mark button at the top of the **Cut-Extrude** PropertyManager to accept the settings and create the feature.

Change the display view to **Trimetric** by selecting **Trimetric** in the View list in the lower left corner of the graphics area.

Select the **Hem** icon in the CommandManager, or pull down the "Insert" menu and select **Sheet Metal – Hem**.

Select the small edge at the bottom right of the base flange as shown. Make sure you select the upper edge line so the hem will be on top of the part.

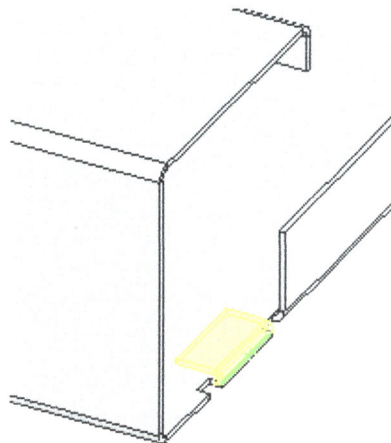

In the **Hem** PropertyManager, select the **Material Inside** button.

Under **Type and Size**, select the **Open** button.

Set the **Length** to '**0.50in**'.

Set the **Gap Distance** to '**0.05in**'.

Select the green check mark button at the top of the **Hem** PropertyManager to accept the settings and create the feature.

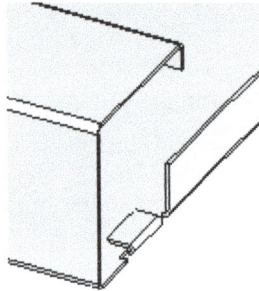

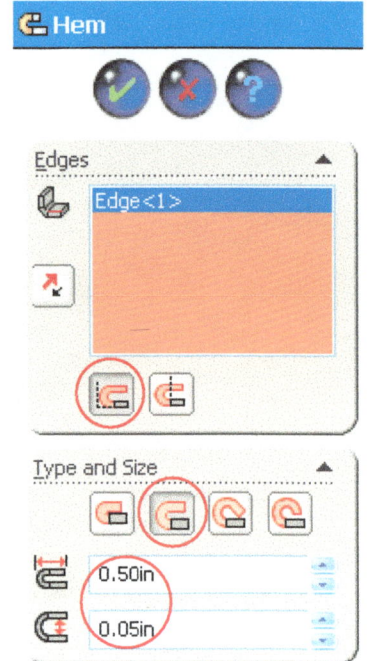

Create a Flange Off the Hem

Select the **Edge Flange** icon in the CommandManager, or pull down the "Insert" menu and select **Sheet Metal – Edge Flange**.

Select the top edge of the hem and move the cursor up and click to set the flange direction.

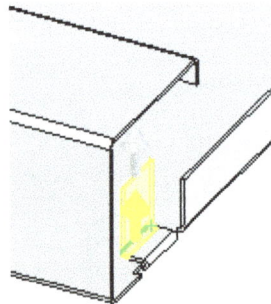

In the **Edge-Flange** PropertyManager, set the **Flange Length** to '**1.00in**'.

Select the **Outer Virtual Sharp** button.

In **Flange Position**, select the **Material Inside** button.

Select the green check mark button at the top of the **Edge-Flange** PropertyManager to accept the settings and create the feature.

Create a Through Hole

Select the **Simple Hole** icon in the CommandManager, or pull down the "Insert" menu and select **Features – Hole – Simple**.

When prompted in the PropertyManager to select a location on a planer face for the center of the hole, pick the top surface of the small flange you just created.

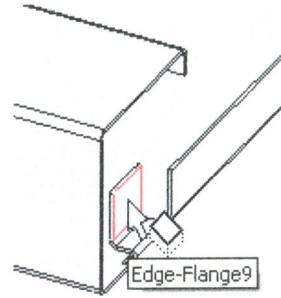

In the **Hole** PropertyManager, enter '**0.136in**' for the **Hole Diameter**.

Check the **Link to thickness** check box.

Select the green check mark button to accept the settings and create the hole.

In the FeatureManager design tree, right click on **Hole2** and select **Edit Sketch**.

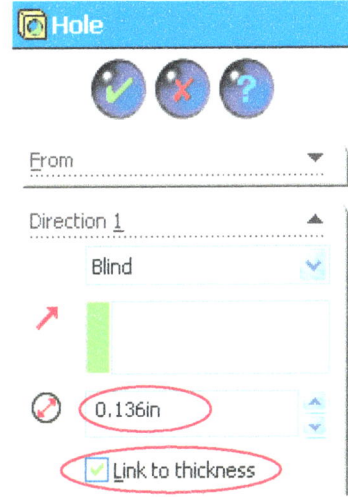

Using the **Centerline** icon, create a vertical centerline connecting the two midpoints as shown.

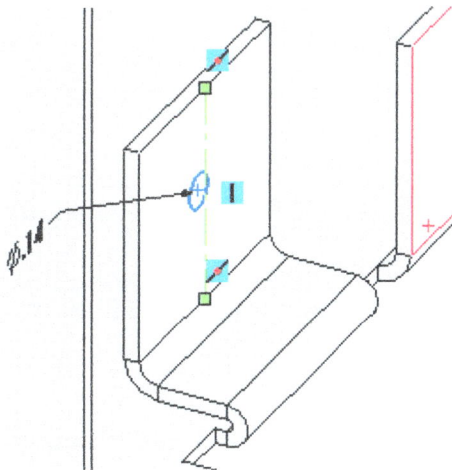

Select the **Centerline** icon in the CommandManager, or pull down the "Tools" menu and select **Sketch Entities – Centerline** to end the command.

With the cursor, left click and drag the center point of the circle to the midpoint of the centerline. An automatic relation will be created, locking in the circle at the midpoint of the centerline.

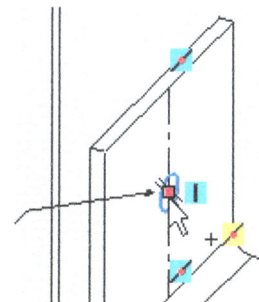

Exit the sketch by selecting the **Exit Sketch** icon in the CommandManager or in the upper right corner of the graphics area.

Mirror the Tab

Pull down the "Insert" menu, and select **Reference Geometry – Plane**.

In the upper left hand corner of the graphics area, click on the plus sign next to the part name to expand the flyout FeatureManager design tree.

Left click on **Top Plane**, and in the graphics area, select the midpoint of the right edge line of the part as shown.

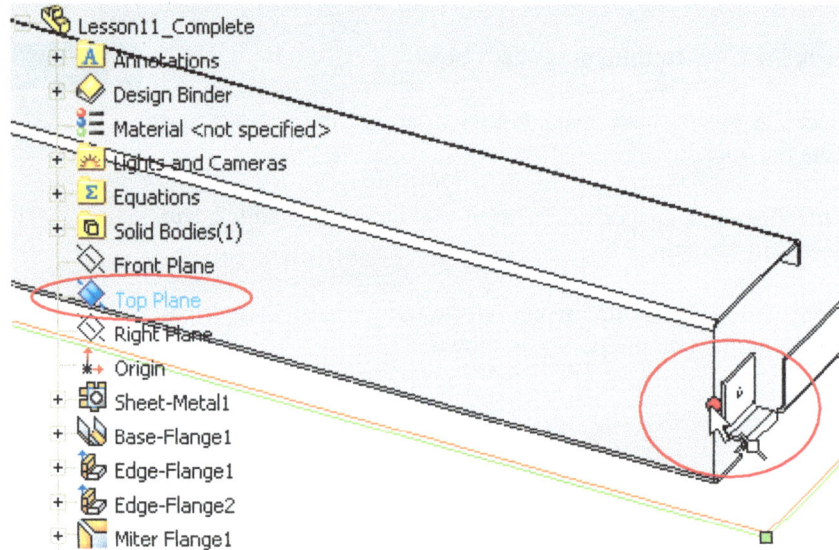

Select the green check mark button at the top of the **Plane** PropertyManager.

Select the **Features** icon in the control area of the CommandManager. Then, select the **Mirror** icon from the toolbar, or pull down the "Insert" menu and select **Pattern/Mirror – Mirror**.

Select **Plane1** as the **Mirror Face/Plane**.

Under **Features to Mirror**, select **Cut-Extrude1**, **Hem1**, **Edge-Flange9**, and **Hole2** from the flyout FeatureManager design tree.

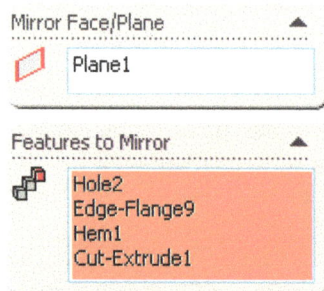

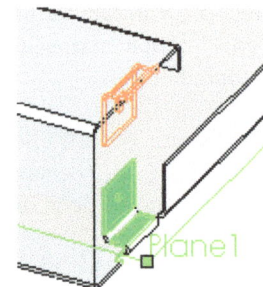

Select the green check mark button at the top of the **Mirror** PropertyManager to accept the settings and create the feature.

The mirrored hem is not correct. Since the bottom hem had auto relief from the edge flange, it is a different size than the mirrored hem. You must go back and add a manual relief.

In order to fix this, roll back the FeatureManager design tree by pressing and holding the left mouse button and dragging the bar at the bottom above the **Plane1** feature as shown.

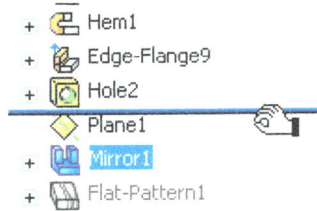

With the part in rollback mode, select the **Sheet Metal** icon in the control area of the CommandManager. Then, select the **Extruded Cut** icon in the CommandManager, or pull down the "Insert" menu and select **Cut – Extrude**.

Select the top of **Miter Flange2** for the sketch plane as shown.

Change the display view to **Bottom** by selecting **Bottom** in the View list in the lower left corner of the graphics area.

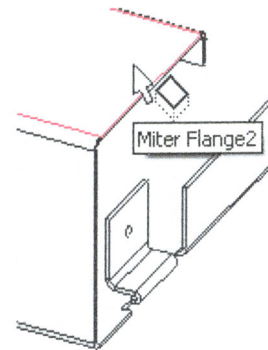

Select the **Rectangle** icon and in the upper right hand corner, create a rectangle on top of the auto relief by selecting the upper left corner and then the lower right corner as shown.

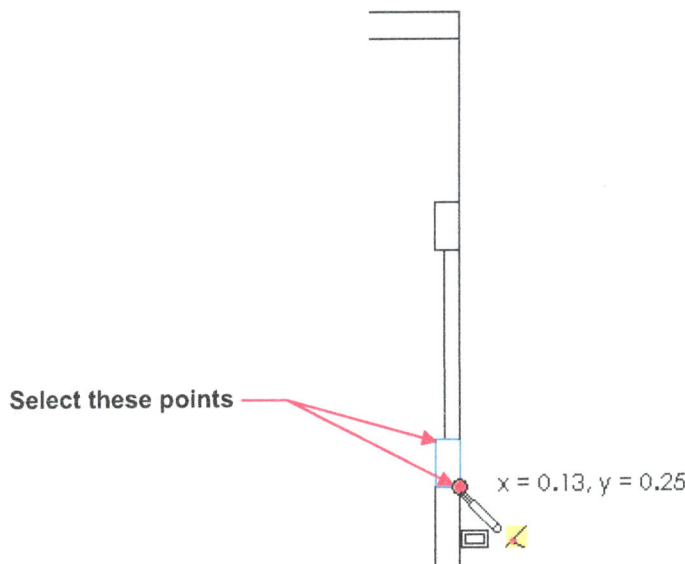

Select these points

x = 0.13, y = 0.25

Exit the sketch by selecting the **Exit Sketch** icon in the CommandManager or in the upper right corner of the graphics area.

In the **Cut-Extrude** Property Manager, check **the Link to thickness** check box.

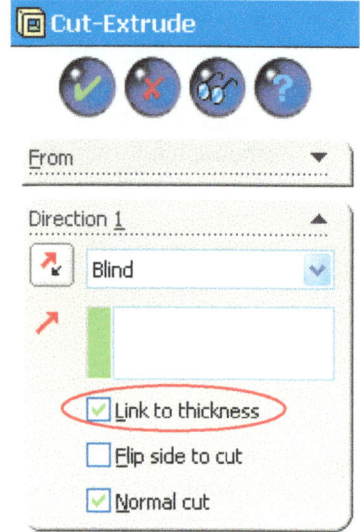

Select the green check mark button at the top of the **Cut-Extrude** PropertyManager to accept the settings and create the feature.

Change the display view to **Trimetric** by selecting **Trimetric** in the View list in the lower left corner of the graphics area.

In the FeatureManager design tree, right click on **Cut-Extrude2** and select **Roll to End**.

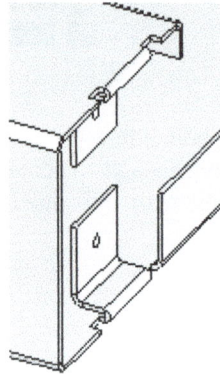

Right click on **Plane1** in the FeatureManager design tree and select **Hide** to hide the plane.

Mirroring the Two Tabs

Pull down the "Insert" menu, and select **Reference Geometry – Plane**.

In the upper left hand corner of the graphics area in the flyout FeatureManager design tree, left click on **Front Plane**.

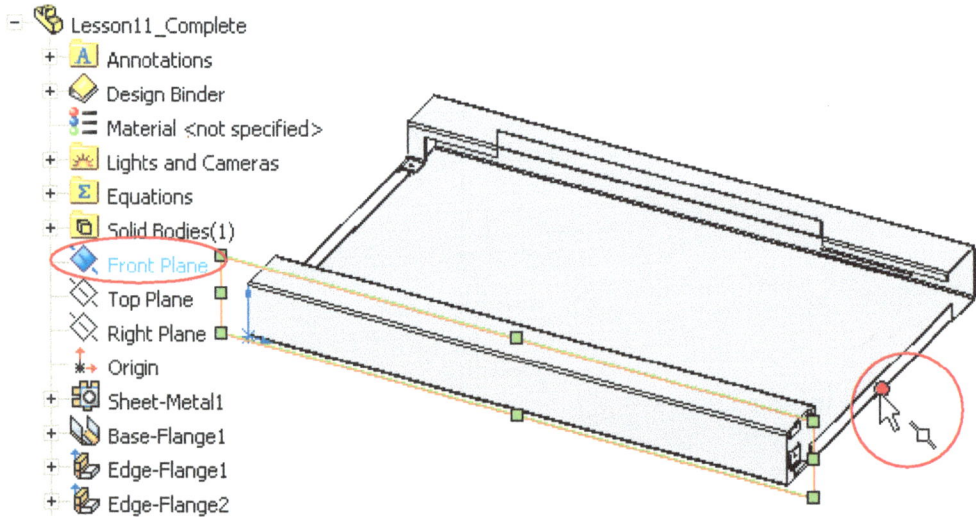

In the graphics area, select the midpoint of the right edge line of the part as shown.

Select the green check mark button at the top of the **Plane** PropertyManager.

Select the **Features** icon in the control area of the CommandManager. Then, select the **Mirror** icon from the toolbar, or pull down the "Insert" menu and select **Pattern/Mirror – Mirror**.

Select **Plane2** as the **Mirror Face/Plane**.

Under **Features to Mirror**, select **Mirror1** and **Cut-Extrude2** from the flyout FeatureManager design tree.

Select the green check mark button at the top of the **Mirror** PropertyManager to accept the settings and create the feature.

Right click on **Plane2** in the FeatureManager design tree and select **Hide** to hide the plane.

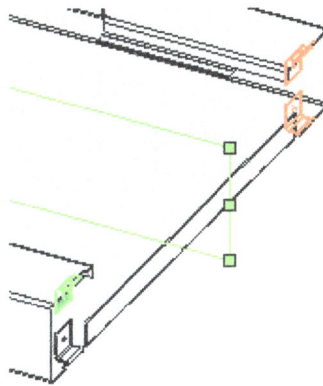

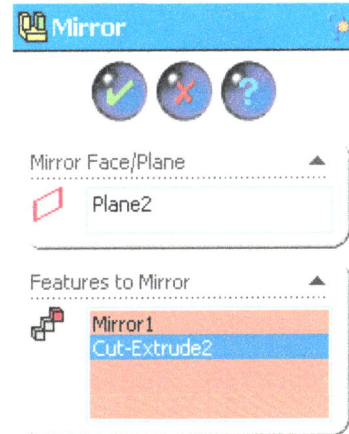

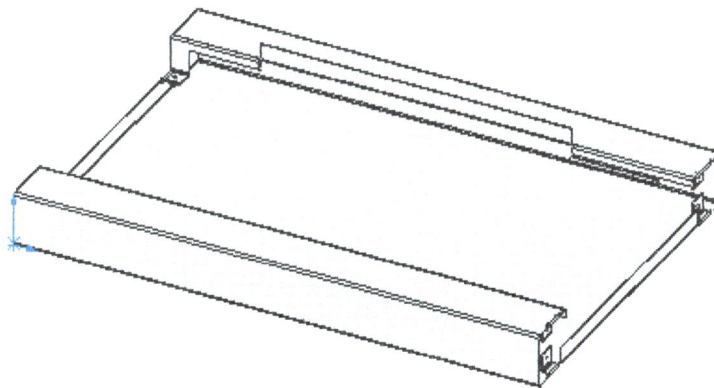

Add Corner Breaks

Select the **Sheet Metal** icon in the control area of the CommandManager. Then, select the **Break-Corner/Corner-Trim** icon in the CommandManager, or pull down the "Insert" menu and select **Sheet Metal – Break Corner**.

Select the corner edges of the all the tabs as shown. Selecting the face of the flange will select both corners.

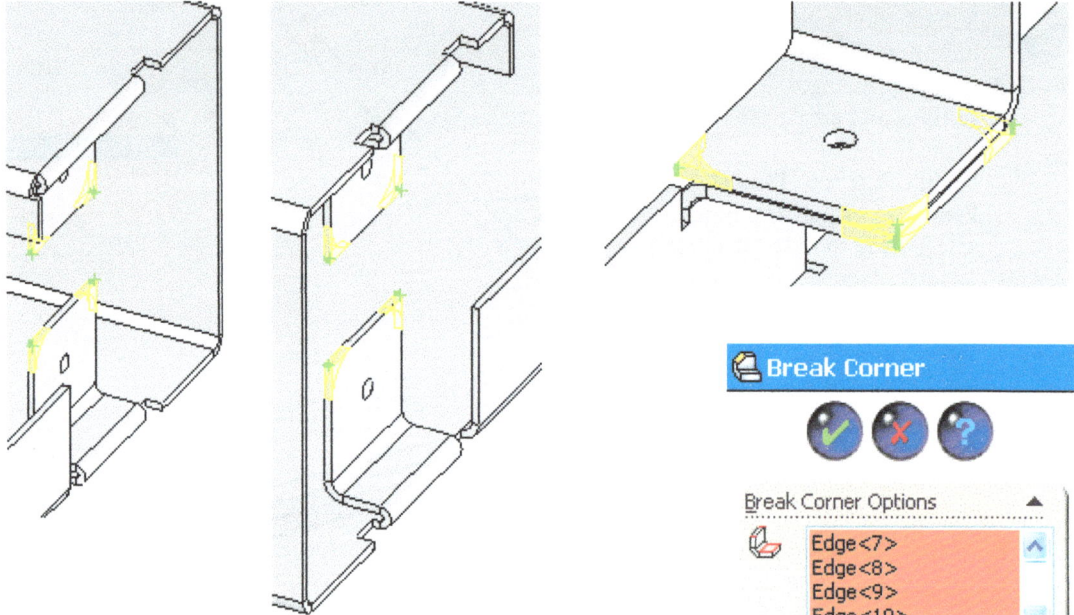

Select the **Fillet Break type** button and set the **Radius** to '**0.25in**'.

Select the green check mark button at the top of the **Break Corner** PropertyManager.

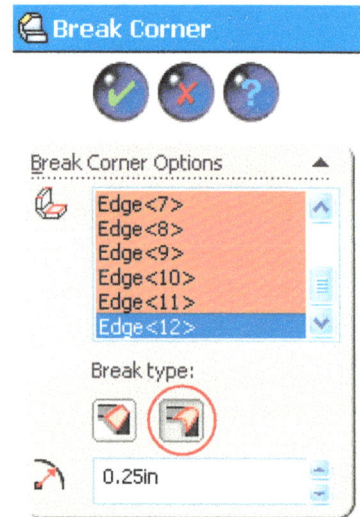

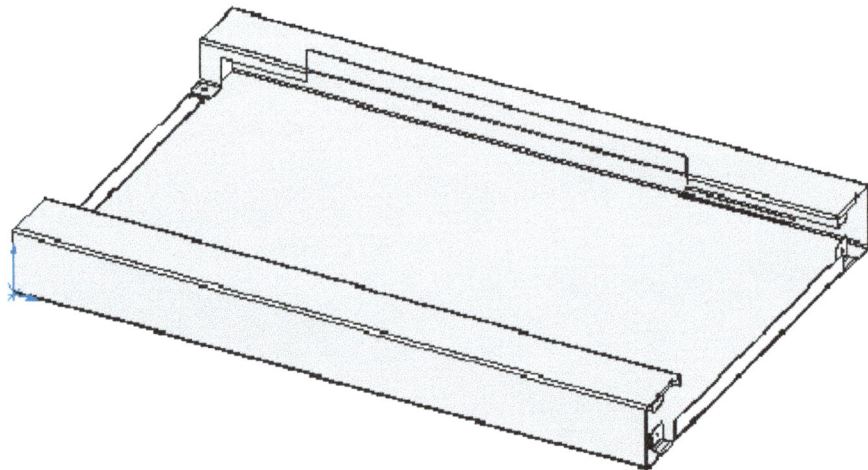

Create a Tab

To do this, select the **Sheet Metal** icon in the control area of the CommandManager. Then, select the **Base-Flange/Tab** icon from the toolbar, or pull down the "Insert" menu and select **Sheet Metal – Base Flange**.

When prompted to select a plane or planar face on which to sketch the feature cross section, select the face of the edge flange as shown.

Change the display view to **Normal To** the sketch by selecting **Normal To** in the View list in the lower left corner of the graphics area.

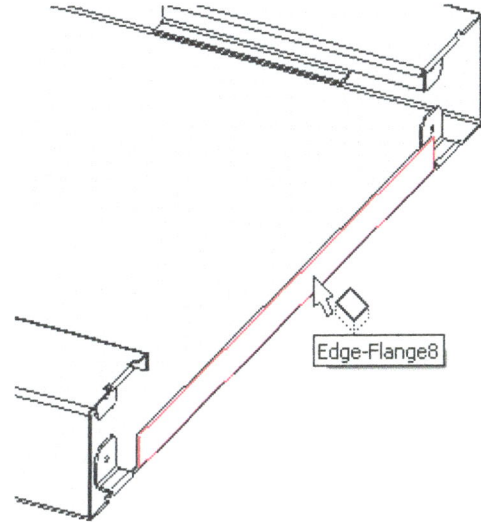

Using the **Line** tool, create four lines as shown. Make sure that you create the bottom horizontal line to complete the closed profile. Do not make the lower right end point of the lines at the Midpoint of the existing horizontal line.

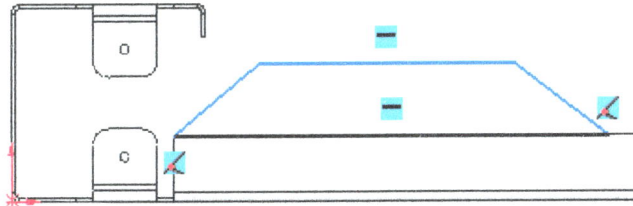

Select the **Add Relation** icon in the CommandManager, or pull down the "Tools" menu and select **Relations – Add**.

Select the two angled lines and then select the **Equal** button.

Select the **Smart Dimension** icon in the CommandManager, or pull down the "Tools" menu and select **Dimensions – Smart** and add the dimensions as shown below.

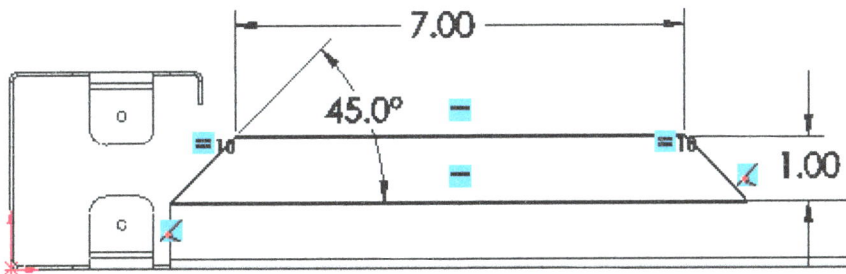

Exit the sketch by selecting the **Exit Sketch** icon in the CommandManager or in the upper right corner of the graphics area.

Change the display view to **Trimetric** by selecting **Trimetric** in the View list
in the lower left corner of the graphics area.

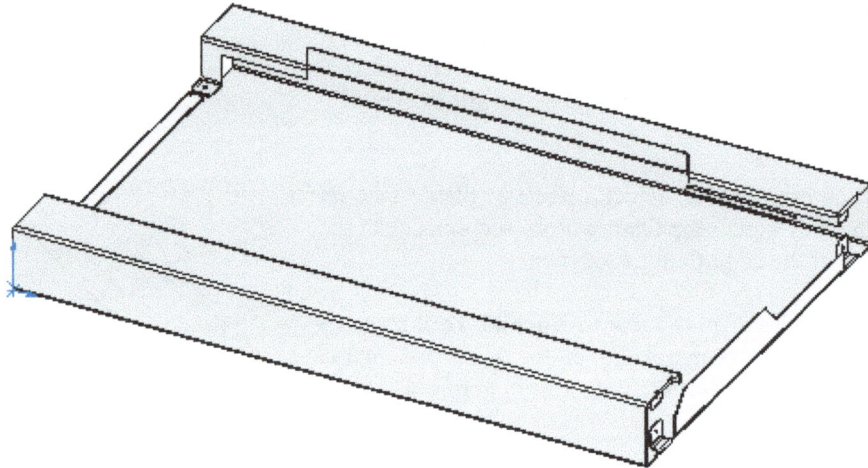

Saving the Part

Select the **Save** icon in the "Standard" toolbar, or select **Save** from the "File" pull down menu.

In the **File name** box of the **Save As** dialog box, type the name of the drawing number. For this lesson, use '**Rack Panel**' and select **Save**.

Congratulations

You have now completed Course 1. Hopefully you have learned how to use the SolidWorks' flange commands to create sheet metal parts. "SolidWorks for the Sheet Metal Guy – Course 2: Hole Patterns and Notches" will take you to the next level. Learn how to create different hole shapes, patterns, and notches on your sheet metal parts.

Index